GASTON BACHELARD

AN ELEMENTAL REVERIE
ON THE WORLD'S STUFF

The Bachelard Translation Series

Joanne H. Stroud, Ph.D., *Executive Director*
Robert S. Dupree, Ph.D., *Translation Editor*

WATER AND DREAMS: AN ESSAY ON THE IMAGINATION OF MATTER

AIR AND DREAMS: AN ESSAY ON THE IMAGINATION OF MOVEMENT

THE RIGHT TO DREAM

THE FLAME OF A CANDLE

FRAGMENTS OF A POETICS OF FIRE

LAUTRÉAMONT

EARTH AND REVERIES OF WILL: AN ESSAY ON THE IMAGINATION OF MATTER

EARTH AND REVERIES OF REPOSE: AN ESSAY ON IMAGES OF INTERIORITY

Gaston Bachelard

An Elemental Reverie
on the World's Stuff

Joanne H. Stroud

Introduction by Robert Sardello

10 digit–ISBN 0-911005-54-4
13 digit–ISBN 978-0-911005-54-7

Library of Congress Control Number: 2014956825

Grateful acknowledgement is made for permission to print the following:
"Prayers and Sayings of the Mad Farmer" by Wendell Berry, from *New Collected
Poems*, used by permission of Counterpoint. © 2012

Natal Horoscope of Gaston Bachelard: 26 June 1884. Used with
the permission of Angela Voss and Jean Lall, editors, *The Imaginal Cosmos:
Astrology, Divination, and the Sacred* (Canterbury: The University
of Kent, 2007), 119. © 2007 by the University of Kent

"Up Against the Wind" by Margot McClean. 17" x 11 1/2", mixed media on rag and
rice paper. Cover image used with permission of the artist. © 2004

Cover design by O! Suzanna Brown
Book design by Eva Casey and Lee Nichol

The Dallas Institute of Humanities and Culture Publications publishes books that bring
disciplines of the imagination—depth psychology, literary criticism, art, architecture,
cultural criticism—to focus on the revitalization of culture.

The Dallas Institute of Humanities and Culture
2719 Routh Street, Dallas, TX 75201
www.dallasinstitute.org

Contents

∽∾

INTRODUCTION

If you are not familiar with Gaston Bachelard and are expecting that I would begin by giving his résumé, instead I ask that you try and enter something completely fresh, new, unknown. And even if unknown, as you will see, you more than likely did not enter the true unknown, but were proceeding tied to the rope of knowing—biographical, historical, philosophical. Many are the possibilities that obscure us from pure innocence.

Formative innocence—that is the qualitative feeling of this wondrous book by Dr. Joanne Stroud. I have known Dr. Stroud for over thirty-five years and have had the extreme privilege of witnessing the unfolding of her intimate love-engagement with Gaston Bachelard. As a close reader of poems, one who receives and holds the image-ideas from poetry as a spiritual guidance in life, no one is more suited to present the vision of Gaston Bachelard, the true master of image-dwelling as life-awakening. It is a delicate task, indeed, for in this time of "spectator consciousness"— looking *at* everything for a mere second of interest and then moving on to the next thing that will capture our attention—few have developed the capacity to linger within the perspective of soul. Such a capacity is paramount for a true understanding of the great imagination, and work of imagination, of Gaston Bachelard. Nothing would be more unsuitable than a cursory speaking "about" rather than what Joanne Stroud presents here—a speaking from "within." It is really that Bachelard has, for Dr. Stroud, become a Way, a Path of the soul and of the heart, the Way of Imagination and the Way of Feeling.

We are thus, by the very form of her writing, being asked to enter into regions that are quite unfamiliar within the world-form of our time—focused as it is on abstract intellectuality combined with cold and brutal imposition of will, all nicely wrapped in superficiality. The middle realm of the heart is all but excluded, given only a small niche within the now-captured region of "the arts," which are "nice" for entertainment, and maybe for possession and display, but feeling is hardly depicted in the world as a way of life. But here, in this amazing writing, we are not only gently introduced into such a Way, it also turns out that nothing—nothing—could be more practical!

Situated within the realm of the heart, the realm of feeling, as Dr. Stroud is in this writing, neither thinking nor will are excluded. From the place of the heart, thinking becomes the "emerging" of thinking rather than the repetition of already-thought thoughts. And will becomes the active "emerging" of true creativity—will as creative-receptive force rather than aggression. This writing is thus a portal, an entry and guidance through the conscious soul realm of pure receptivity and the action of receptivity, with its untold power and unlimited dimension. This book is nothing less than a manual for discovering our long lost capacity for true joy! Like all capacities in the act of forming, patience, concentration, yielding, allowing, going over what seems already covered, discovering that it is never the "facts" but the nuances that are significant, are the necessary practices that are built into this writing—practices that are already the Way. We get to the "goal" by being already, each moment, within it.

Quite directly said, Dr. Stroud presents Gaston Bachelard in an exquisitely "Bachelardian" fashion, through images, in images, and through what we might think of as "the great reversal"—being more "objective" about our inner life while at the same time becoming more "subjective" in our presence with the world. We discover that it is quite possible to carefully and with care "track" the inner world through deepest engagement, and at the same time feel everything around us as a Who, as Presences. The imaginal world discovered by Gaston Bachlard, lover of poetic images, that great alchemical union of "knowing and presence," can heal the world, can heal the soul, can heal our relationships, from the most intimate pairing to the boardroom, even leading to international intimacy of respect. This book presents these possibilities.

We live in a quite strange, certainly unique world that can best be described as a "me cosmology." It is not a philosophical stance per se, but it is the way we locate meaning, and it is also the way world destruction now operates—self-interest as the only interest—and indeed, the very notion that there is something else cannot be seen. Even religion and spirituality have fallen into this world-form, for they are now both completely oriented toward how they can bolster our ego-subjectivity as all-important. This characterization intends to be descriptive rather than evaluative. I feel it is necessary to uncover the main assumption that holds us captive, for this book speaks a different cosmology.

In the past, the greater cosmology, one uniting earth-world-cosmos-individual would have been called "alchemical cosmology." In more directly experiential terms, this cosmology holds that images are real, are primary, are formative of the ways we sense and perceive and act—because this is the way the universe itself knows itself. Images, in their perpetual act of forming, are simultaneously "subjective and objective." At least so for true images—conveyed to us through poetry, so beautifully selected and integrated into this writing, amplifying what Bachelard himself engaged in.

A formative unity exists, exemplifying the dynamic earth, now and forever undergoing and guiding its own creating through the elemental imagining forces of Earth – Water – Fire – Air. These are the primary images engaged by Bachelard and also by Dr. Stroud. They are and are not the elements we experience—physical earth, water, air, fire. They are what we see, feel, sense, rely on completely—and they are also subjective, active and real presences, known in their immediacy, forgotten in their subtly, brought out and re-introduced by Bachelard and Stroud, promising something that can be delivered—life, home, truth, presence, world-intimacy.

Even though Gaston Bachelard indicates an indebtedness to C.G. Jung and the notion of archetypes and archetypal imagination, that association can take us away from the immediacy of discovering how we learn about ourselves through world revelation. The path "inward" cannot, without dire consequences, be divided from the path "outward." But it is not a matter of switching willy-nilly between the two. The privileged domain of images, and in particular elemental images and images from poetry, not only hold the two domains together—they are immediate evidence that separation itself occurs secondarily and comes about through abandoning the heart for the necessarily digitalizing mind. This truth is not presented as a thesis by Dr. Stroud, but it is enacted through *intensification*—making this truth a presence rather than a pronouncement.

Dr. Stroud has been about this task of "grasping, then mapping, and then making something entirely new" for some forty-five years—a way of interior development given through her most important mentor, Donald Cowan, university president, physicist-poet. Grasping is not a greedy, pretending-one-is-a-master but the primary act of the open and receptive heart. Dr. Stroud reads Gaston Bachelard in this manner and invites us to linger there.

Since little of Bachelard's work was available in English translation thirty-five years ago, Dr. Stroud had to have translations made of his major works and make them available. That alone would suffice for a life's work. In so many ways, however, it was the "tool-gathering" necessary to then inwardly allow the material itself to form its own territory within the soul—a long and arduous "mapping" of the entirely unknown regions of conscious depth imagination.

A significant departure from Jung, initiated by Bachelard and carried forth by Joanne Stroud, is of great significance, and points toward the third dimension—"making something entirely new." It is the difference between working, as Jung does, with a kind of schema—grand, for

sure, and working phenomenologically, as Bachelard does—
letting the matter itself reveal itself on its own terms. Such
an approach involves a loving self-effacement. Dr. Stroud
presents world-speaking images rather than what we think
we know about the world. But, even more so, valuing the
"imaginal-real" as present not only in our night-dreams but
in our day-dreaming and in joyful flights of reverie, and in
sensing and perceiving too—that is what I mean by "loving
self-effacement." Said more positively, it is the creative act
of making something entirely new. Here, imagination is
utilized to intensify imagination, and intensified imagination
becomes the organ for perceiving the imaginal worlds—
as primary, as originating, for both Earth and the Human
Being. Few are able to bear caring for something more than
oneself.

    With Bachelard, and with Dr. Stroud, we experience
something of a revolutionary shift: that it is indeed possible
to live poetic, imaginal existence as formative of daily life.
Bachelard seemed to be somewhat of a recluse, which seemed
necessary to the work he was doing, for the disturbances
of the hectic world would have delayed or even foiled the
unfolding of his work. Dr. Stroud, however, intimates that it
is possible, indeed imperative in this time, to be fully within
the world as active world-dreamers. While the process may
well be the "grasping, mapping, making," equally important
is the primacy of the imaginal Earth, and dedication to
imaginal ecology.

    I am speaking of Earth as more than one of the
four formative elements. And certainly more than our
conceptions of Earth as a ball of earth strewn with plants,
animals, water, the source of all our resources. One simply

cannot live such a removed conception of Earth after reading this book. Not only is Earth, perceived through the poeticizing reality of imagination, a being of body, soul, and spirit—some aspects of which are visible, some which are not—but this being is a She, and She can only be known through engagement according to the laws of imagination. All of Earth speaks imaginally and emotionally. The little manual for incorporating this "Way of the Earth" rests within your hands. Perhaps the most poignantly true words of the whole of this writing are when Dr. Stroud says— "Taking Bachelard to heart, it would become anathema to mutilate our precious planet."

It is both surprising and disheartening that Bachelard is not better known. It is equally surprising that his reputation rests primarily within the academic world, though he has had some influence on artists and architects. But the wider world has not heard of him, in spite of the fact that Bachelard himself remained close to the world of physical work during the whole of his own teaching career. Within this book, Dr. Stroud has perhaps ignited a larger imagination for Bachelard. She does not do so by trying to promote Bachelard, but rather by being true to his vision, and his method. This book is densely packed with pregnant hints of how to proceed.

The act of attention must be considered in relation to image-presence. Dr. Stroud speaks of the "arousal of attention"—the force of attraction, coming from the image itself, that ignites our interest. The great mystery of attention begins with wonder concerning its nature and then noticing its ability of attraction. When attention, for example, is centered in fear, as everyone's is in this time, that attention attracts, amplifies, and multiplies fear. We cannot rid ourselves of fear by instituting procedures intended solely for protection—from firearms to perpetual surveillance. It is freedom of spirit that opens totally different attention, attracting then totally different surrounds and action. Freed attention is a primary message of this work on Gaston Bachelard, and poetic images are the holders, with open hands, the very essence of freedom of spirit.

The form of freed imagination promoted by Bachelard and here so strongly reinforced by Dr. Stroud, who clearly perceives the need of the times, is active imagination. The world is currently overtaken with images, none of which inspire freedom because they obscure rather than ignite the spirit. Passively taking in images captures, rather than frees, attention. We come then, in this book, to see the world-therapeutic value of the active imagination. Such world-therapy begins with actively noticing what attracts our attention. This small act of noticing is a huge, endless space/time of freedom being born.

Poetic images perhaps hold the privileged place of "throwing" us into discomfort, throwing us out of sleepiness just enough to arouse attention. Then, at that exact moment, a second motion of attention is required—actively ourselves to be carried inwardly by the utter newness of what has offered itself to our unknown freedom. This motion of being inwardly carried into the new, is reverie, dreaming while awake. These two motions—arousal of attention and yielding to dreaming, are, however, not passive processes but *receptive actions* that are in complete resonance with the essence of true freedom. These motions which form nearly all of the work of Bachelard, and which he does describe, in his own way—now, through the way that Dr. Stroud proceeds, become even more consciously available, shifting the fulcrum of Bachelard from the beauty of world dreaming to the magnificence of the therapeutic action of the inherent poeticizing of world-presence.

How little is asked to change the whole world! Nothing more than careful and caring attentiveness, held within the soul and allowed to dream itself through us while we remain awake. How available the path through constant anxiety is! Big acts of creating are neither necessary nor needed, but creating is necessary in order to go through the constant rhythm of expansion and contraction. Simple presence to the inherent creative action of the will of the soul unknots anxiety at the very moment we come close to stopping breathing. Neither Bachelard nor Dr. Stroud are suggesting that all poetic images have to be beautiful for us to find freedom. Images of aversion are as important for the attention-reverie enactment as those of desire. Both together keep the psyche from falling into stupor and oblivion.

Psyche is also a focus of attention throughout Bachelard, and even more so in this book. However, we are fortunately not led into simplistic psychologizing, reducing all experience to one's personal psychology. The distinction between "human-nature" and "person-nature" must be made and emphasized. Every word within this book, and within Bachelard, has to be understood as being a phenomenology of "human-nature." Such a concentration is inherently therapeutic because the psyche is central to human-nature, particularly the human ability of existing within soul-awareness.

Psychology, as we now know it, does not have such a focus. The focus is rather on "person-nature," and if anything, the needed understanding of human nature upon which to form an approach to person-nature is completely bypassed. That modern psychology has betrayed its calling and at the same time has become a huge shaping force for our understanding of humanity inevitably means that at first this book will seem strangely attracting and strangely odd. Quite naturally, a reader will tend to read what is written here in terms of "person-nature"—that is, what is said of Bachelard will be automatically understood as if it is a kind of odd psychology, and at the same time the feeling will be present that something much, much more significant is proposed. While, as Dr. Stroud points out, images are psychic realities, the psyche exists quite independently of acting as if it is our possession. And psychic realities are also quite independent of our person-ality, though nothing is more intimate. The mind balks at the contradiction while the soul sighs in relief.

Consternation can be somewhat alleviated by a close reading of this book, allowing yourself plenty of moments of reverie while doing so. Notice, particularly, the way in which Dr. Stroud carefully leads us into the realization that images are not visual, that imagining does not mean "having inner pictures." Further, the very arrangement of this book is itself an instruction in the true fullness of images and imagination; they are not separate from matter, but, as she says, what make matter matter. Images are also not personally subjective, but they are the resonating inherent subjectivity of human nature. To be human is to be subjective, but not person(ality)-subjective—only modern psychology subscribes to the latter. And as inherent within matter, images are bodied and "worlded" simultaneously. The problem with psychology is that it has taken the "registering" of what inwardly occurs to be the same as what inwardly occurs. Thus, psychology is inherently ego-bound. There are only two ways around such confusion. Jung takes one: symbolic imagination always bypasses ego, but requires the insertion of interpretation; and phenomenology, allowing the phenomena to reveal itself, the way of Bachelard, and now of Stroud, which relies on the privileged status of poetic images.

If we try to implicate Bachelard in psychology, then there will also be the accompanying disappointment of the absence of technique. There is nothing within Bachelard, nor within Dr. Stroud, that hints that there is some way of arranging circumstances to bring about inner change, as is the purpose of technique. The technique of free association, for example, allowed Freud to interpret the person dimension of the unconscious. The technique of archetypal amplification allowed Jung to bring personal and collective consciousness into a harmonious whole. We may be tempted to think that attention to poetic images and reverie are the "techniques" of Bachlard. But they are his *method* rather than techniques. Method is an intensification of the Whole, through an act of attention, that makes the Whole present in an instant of particularity. Dr. Stroud inherently understands this. Everything she writes is of individual interest, concern, and attraction while at the same time her writing forms a way of speaking within Wholeness, revealing that she proceeds, as does Bachelard, by means of method rather than technique.

How, then, does the method of Bachelard work? This is the ongoing sub-text of Dr. Stroud's writing—she is indeed revealing not only his method but specifying exactly the way that it works and thus can work for anyone—for he is speaking of the function of imagination as central to human nature. Within such a stance, there is nowhere to "get to," no development that needs to occur, for it is all already present. It is not that some have developed to the point of being present to images while others have not, and there is a set of techniques that can help development. There is only attention—and its awakening. There may be "levels" of attention, considered as something like levels of vibration. But such levels are not of the nature that makes one have something while others do not, and those that do not are "abnormal." Within the cosmology of Wholeness, there is equality of spirit, inherently. The difference between our presence with images, and a poet's, is a matter of the expansion of Wholeness rather than development of a technique.

Expanding image-presence does something magical—it re-organizes the will. Dr. Stroud has followed this line in her previous book, *The Bonding of Will and Desire*. The insights of that book are carried forth in this book in exciting new ways. Through image-presence, the inherent attraction within image, the presence of desire as autonomous and inherent—as central to human nature—is transformed into the will-to-action. This transformation occurs because through the elements of matter, Earth – Water – Air – Fire, we come up against something so completely "Other" and so completely "Us" that world-revelation occurs, and this revelation does not sit there as mental insight but mobilizes individual will into "world-will."

INTRODUCTION

The "method" of Wholeness does not, however, proceed by logic, nor does it follow the laws of cause and effect, both reasons why Wholeness is outside the rightful province of technique. Imaginal Wholeness follows the laws of association, of felt-closeness. Thus, it can appear to the mind needing logic and conclusions that imagination is equivalent to disorganization and the law of association appears as "free associating." The writing of Bachelard—and also of Dr. Stroud, as an initiate of an initiate—follow the laws of Wholeness. They are consistent in doing so, and you will notice that the habit of the mind to sweep along, making instant conclusions of whatever it encounters, is completely thwarted when coming up against this writing. At the same time, the will to "stay-with-it" has an amazing result. You are changed in your vitalness, in your vitality. The world actually becomes sensorial radiance; so do other people; so do things. And, no matter what the challenges of life might be, they now re-organize according to a different principle—that of the revelations of the Wholeness of individual-Earth, an awakening to the Spiritual Earth.

We do not have to choose between "organization" and "life." There is an important linking that can occur that resolves the choice and makes both possible as a new simultaneity of life's order. This linking is the vast presence of Silence that is felt, almost as "sweeping in," when dwelling within the poetic image. You can feel this Silence permeating the writing of The Seduction of Matter. If you at first feel a kind of restlessness, maybe even a frustration, these are the two signs of being at the outer edges of the vast presence of Silence, the organizing presence with the phenomeno-logic of images. It is the inherent Silence, the utter Stillness that is the power of imagination—and the power of this book.

Robert Sardello

PREFACE

My first contact with Gaston Bachelard was somewhat comparable to a first love. I was smitten from the moment I read his *The Poetics of Space* as a graduate student at the University of Dallas during the 1970s. I knew that we were made for each other. I didn't quite realize then that it would be such a demanding relationship, but forty years later my heart can still throb in response to Bachelard's summons to be alert to the nuances in nature, his call to celebrate the earth's beauty and the magnificence of our everyday surroundings. His lyrical voice inspired me, enriching and deepening my appreciation of the world around me. I was prompted to explore some of his more complex theories of the functioning of consciousness and especially the imagination. Bachelard will not allow the sidelining of the imagination. He considers it neither peripheral nor a special gift more available to a select few creative people. It is not just an abstract force that we can ignore. It is not merely that which illuminates or darkens our experience. Instead, Bachelard summons the imagination, revealing as he proceeds how it underlies and informs the world we perceive. Bachelard radically changes the way we understand imagination. For Bachelard, the imagination is central. It dominates and determines how we see the world. I agreed.

While nearing the end of my work toward a Ph.D., I was given my first job teaching a freshman core curriculum course in the genre series of the U.D. Literary Department, headed by the teacher I most admired, Dr. Louise Cowan. I couldn't make up my mind if I was a psychologist with a love of poetry, or a literature teacher with a framework of psychology. Both disciplines had interested me all the way through my undergraduate years at Sarah Lawrence College, when I avidly studied Carl G. Jung with Joseph Campbell and had, as my don, Marc Slonim from the Russian Literature Department. In the 1970s, I was growing discontented with the field of literary criticism with all the picayune analysis, with its emphasis on deconstruction that was current at the time. It seemed to me that all the pettiness of poetry analysis was just so much of nothing and was, in fact, a way designed to lose touch with the poem itself. I decided that I might have much more to give to the area of psychology, so I switched my major to psychology where another special friend, Robert Sardello, headed the department of Phenomenological Psychology. That is a long name for studying how the world around us is subjectively experienced. While working on an interdisciplinary degree, a Ph.D. in Psychology and Literature, I experienced some trouble convincing my dissertation readers of how to combine effectively my work in Jungian psychology with my interest in twentieth-century poetry.

When I found Bachelard, it seemed like a "marriage of true minds," to borrow Shakespeare's line. I couldn't help but be intrigued by his provocative questions, such as this one, which combined both of my interests in psychology and in poetry: "How can an image, at times very unusual,

appear to be a concentration of the entire psyche? How—
with no preparation—can this singular, short-lived event
constituted by the appearance of an unusual poetic image,
react on other minds and in other hearts, despite all the
barriers of common sense?"[1] Here was a thinker who
could also describe the feeling of elation when writer and
reader communicate. How unusual is Bachelard's direction
concerning the demand required of the reader in order
to capture the essence of a poem: "The reader of poems
is asked to consider an image not as an object and even
less as the substitute for an object, but to seize its specific
reality. For this, the act of creative consciousness must be
systematically associated with the most fleeting product of
that consciousness, the poetic image. At the level of the
poetic image, the duality of subject and object is iridescent,
shimmering, unceasingly active in its inversions."[2]

Bachelard is here addressing a fundamental question of
how consciousness works. He is also touching on another
aspect close to my interest in Jungian psychology. How can
different minds, with diverse backgrounds and distinct
cultures, across the expanse of oceans, react to and be
stirred by the same images? What human trait is responsible
for the collapse of differences when minds are joined? Jung
attributed this phenomena to the "collective unconscious,"
common to all; Bachelard to the imagination shared by all.

Etienne Gilson, in his foreword to *The Poetics of
Space*, explains why Bachelard's concept of imagination is
hard to comprehend: "What Bachelard calls imagination
is a most secret power that is as much of a cosmic force
as of a psychological faculty."[3] For Bachelard, imagination
functions through classical forms in the way philosophers

xvii

have always delineated, creating beauty. But, in addition, Bachelard's special contribution is in the area not of formal imagination, which he considers important, but rather in another function of imagination, which he calls "material imagination." He was aware that he was doing pioneering work in turning to "images of matter." This switch in his interest from the precision of the Sorbonne philosopher of science to the study of poetry marked a major reorientation of his past education and his finely-honed *modus operandi*. Here is how Bachelard himself explains the effort—one that required "a minor daily crisis":

> A philosopher who has evolved his entire thinking from the fundamental themes of the philosophy of science, and followed the main line of the active, growing rationalism of contemporary science as closely as he could, must forget his learning and break with all his habits of philosophical research, if he wants to study problems posed by the poetic imagination. . . . One must be receptive, receptive to the image at the moment it appears: if there be a philosophy of poetry, it must appear and re-appear through a significant verse, in total adherence to an isolated image; to be exact, in the very ecstasy of the newness of the image. The poetic image is a sudden salience on the surface of the psyche.[4]

Actually, though, Bachelard's two sides don't seem as incompatible to me as to many others. After that major switch in 1938, in which he shocked his readers by psychologizing the element of fire, he continued to shift back and forth in his writings, exploring studies in both science and in poetic imagination. I maintain that we cannot disassociate these two sides of his work—the phenomenologist, the explorer of imagination, and the explorer of scientific theory—studies normally requiring diametrically opposite approaches. The very detailed, analytical work in science, I believe, furthered his acute powers of observation of the physical world that undergirded his interest in images of matter. The integrating force of the imagination, which he attests to over and over again, is at work in both his areas of concern.

I was so intrigued with my first tasting of Bachelard that I asked: What else has he written? I discovered two other books that also had been translated into English, *The Psychoanalysis of Fire* (1938), in which he made his first move into the imaginative realm, and *The Poetics of Reverie: Childhood, Language, and the Cosmos* (1960), which was written shortly before he died in 1962. In addition, there was that marvelous compendium, a collection of excerpts from all his works, *On Poetic Imagination and Reverie*, that Colette Gaudin assembled in 1971, and that Spring Publications published in a new edition in 1987. What I also found was that there were many of his works on the elemental imagination that had never been translated. The alchemist in me, who had always responded to the four classical elements, longed to be able to make these available to English-speaking readers. So, off I boldly went to Paris in 1981 with the intention of tracking down José Corti, one of Bachelard's two publishers.

With some trepidation, I signed a contract for the rights

to translate *Air and Dreams: An Essay on the Imagination of Movement; Water and Dreams: An Essay on the Imagination of Matter; Earth and Reveries of Will: An Essay on the Imagination of Matter*, and its companion book, *Earth and Reveries of Repose: An Essay on Images of Interiority*. I was into the project so deeply that it seemed a shame not to complete the series. So in 1986, I went back to Paris to José Corti to get the rights to *Lautréamont*; I also went to Bachelard's other publisher, Presses Universitaires de France. His work on imagination had begun with the image of fire in *The Psychoanalysis of Fire*. He was still working on the image of fire over twenty years later. With PUF, I contracted to translate *The Flame of a Candle*, and also the work that was unfinished when he died in 1962, *A Poetics of Fire*, which the Dallas Institute published in 1990 as *Fragments of a Poetics of Fire*. In this final book he said: "In my choice to study poetic images of fire I was luckier still, for this involved the study of inflamed speech, reaching beyond all decorative intent, at times even aggressive in its beauty."[5]

I also obtained the right to reprint *The Right to Dream*, a posthumous collection of essays and introductions to the work of many artists. I have been at this involved endeavor of translation ever since. In November 2011, at the Dallas Institute of Humanity and Culture's thirty-year celebration, we completed the Bachelard Translation Series.

It has been said of Bachelard (I am paraphrasing and adding to this statement): Bachelard is a phenomenologist who lets the world speak through him rather than concocting a theory about the world. We are allowed to begin where we are, as ensouled beings, receptive to the currents going in all directions—ever gathering momentum by multiple associations, soaring lyrically, or going downwardly deep. He covers it all, the whole range of psyche. Truly, that's it.

I want to mention what some of his critics and observers have said about his work. Mary McAllester Jones, a major translator of his works, both poetic and scientific, mentions Bachelard's broad, inclusive scope of interest and his refusal to be reductive or dogmatic in any way: "Bachelard constantly expresses his dislike and constant refusal of dogmatism in any shape or form, his intellectual creativity and the wide range of his interests are like water in the desert . . . of modern French thought."[6] Colette Gaudin similarly sees the extensive nature of Bachelard's explorations: "The work of Bachelard stands beyond the disputes of critical theory as a monumental reflection on human knowledge and creativity."[7] Gaudin wants to see Bachelard "take his proper place in the current philosophical debate on human culture and its interpretation." Roch C. Smith stresses the importance Bachelard places on imagination: Bachelard's work demonstrates that "our experience of the world does not govern our imagination and our verbal images; rather it is our imagination and associated verbal images that guide our experience of the world."[8]

Mary Ann Caws prefaces her study of Bachelard as "the philosopher of surrealism" with a quote from his *Paysages d'Albert Flocon* [Landscapes of Albert Flocon]:

> "According to our courage or our lassitude we'll say that the world has begun or that it is finished," along with a quotation from the founder of surrealism, André Breton, in his magnum opus

Surrealism and Painting[9]: "When I know what will be the outcome of the terrible struggle within me between actual experience and possible experience, when I have finally lost all hope of enlarging to huge proportions the hitherto strictly limited field of action of the campaigns I have initiated, when my imagination recoils upon itself and serves only to coincide with my memory, then I will willingly imitate the others and grant myself a few relative satisfactions. I shall then join the ranks of the embroiderers. I shall have forgiven them. But not before!"[10]

As a teacher of many years, I always admire the ability to communicate clearly, another reason for me to identify with him. According to McAllester Jones, "Bachelard was a teacher, and by all accounts, a marvelous one . . . . In his last lecture that he gave at the Sorbonne on January 9, 1955, he said, 'For me living and teaching have been the very same thing.'"[11] I always claim that Bachelard is one of the most incredibly original minds of the twentieth century. I would be pleased if you would be able to agree that this statement is not the natural hyperbole that one can indulge in when spending so much time with a highly sophisticated French gentleman, but is indeed true. I have to stop and remind you that his background was most unusual in France, where academic careers are very structured. His was purposeful, but arose out of a humble beginning.

Bachelard was born on June 27, 1884 in Bar-sur-Aube, in the Champagne country, the son of shop owners. After high school he worked in the postal service from 1903 to 1913, while studying at night to get his bachelor's degree in mathematics. He spent a long five years in the military in World War I and was decorated with the Légion d'honneur. He taught physics and chemistry in secondary school from 1919 to 1930, while continuing his nighttime studying. At the age of 35, he began graduate work. He wrote his doctoral thesis in 1927 and went to the University of Dijon to teach for ten years. By the year 1940, when he was appointed to the Chair of History and Philosophy of Science at the Sorbonne, which he held until 1954, his reputation as a writer on the "new scientific spirit," as he called it, was well established. He remained at the Sorbonne until he became Emeritus in 1955 at the age of 71.

"In the lecture hall he was capable of reaching out to each listener, of capturing attention with his playfulness, geniality, and genuine concern for all those around him. Only *seeming* to ramble as he spoke, he would discover and explore the images that came to him, often stopping to give a rare word its interval, or to create new words that somehow always seemed to have existed before,"[12] is how Gaudin describes him. He looked the part of a philosopher with his bushy white head and beard. He was often seen strolling slowly "with the gait of a farmer at home on his native plot of ground," down the Boulevard St. Germain talking avidly with students. That he achieved such success later in life was, as McAllester Jones emphasizes, quite "outside the normal avenue of competitive examinations taken by bright young French persons,"[13] especially for someone who worked sixty hours per week minimum, which was even more unusual. Peter Caws makes this intriguing connection about Bachelard's working life and its influence on his academic and writing career:

On the whole it seems to me that it would be a good thing for more philosophers to have been postmen. The métier may not be accidental: apart from the letter-scales Bachelard refers to as having given him his idea of weight, there is a hermetic side to the postman's activity—he is the point of contact with the world beyond, he brings sealed messages from distant origins, there is no knowing what marvels or portents they may not contain; at the same time nothing can surprise him, he is the very image of persistence and reliability, of local intimacy and homely order. And when the postman himself leaves for the outside world—for Dijon, for Paris —he takes with him this imperturbable sense of the familiar, and his concern continues to be with the firm materiality of the world, now from the scientific point of view.[14]

In 1984, to celebrate the centenary of Bachelard's birth, a postage stamp was issued in his honor. Postage stamps in honor of a philosopher of science may seem surprising, yet Bachelard was, in addition to being a beloved teacher, a much-admired and influential figure in France, although far less known to English-speaking readers.

What is unique about Bachelard? I think that he has a special talent to bring ordinary objects of the living world into extraordinary focus. His scientific training may have influenced his minute observations and his singular powers of concentration on whatever he viewed around him. This characteristic also makes him difficult to read. His writing is so dense that, if I expect to take it in, I can only read a

PREFACE

few pages at a time. But when I do settle down, stop racing, and let my mind absorb what he is saying, then something very mysterious happens—my mind seems able to juggle everything around and to make new connections. Even unrelated problems seem to find solutions. In short, he puts my imagination into high gear. This practice of slow reading is a major contrast to the short attention span required for so many of our activities in this current time of cursory glances. Jean Lall, rather than complaining about how one has to reduce speed when reading Bachelard, considers this quality a virtue: "Bachelard's prose . . . slows the reader down . . . and positions her close to where the image is flowing or concealing, weeping or singing, so close that it would be impolite to get up and leave. This slowness of reading and the difficulty of summarizing or extracting concepts from the page seem to me to have a 'philosophical' effect. As the craving for information and mastery gives way, the love of wisdom is awakened and one becomes attentive to the obscure vegetation, the black flowers that bloom in the depths of matter."[15] I find Jean Lall's book with her essay on Bachelard a delight (Lall is a Jungian psychologist with a bent for reading horoscopes). She calls our attention to Bachelard's very telling horoscope with its heavy concentration in only three houses. For those interested I refer you to page 144 in the appendix.

A poet friend, Gerald Burns, sadly no longer alive, wrote to me, "He [Bachelard] stays with the image, dwelling on it, talking about it, until he is even a little past the point at which he's in danger of being a bore. It's then, when he has sunk into near-witless repetition and gabble, that the best things well up." Rosmond Bernier, writer and lecturer on art history for many years at the Metropolitan Museum in New York City, personally knew Bachelard, and took one of his courses at the Sorbonne. He spent the whole semester talking about a single image, *le trou*, "the hole." She agrees that Bachelard is difficult to capture in translation. He has a soaring lyrical voice, somewhat akin to the romantic poets. He is in love with the things of this world. In his explorations of matter, what he calls "material imagination," he brings originality to whatever and wherever he casts his glance.

Although his images are primarily material images, Bachelard explores many images that Jung or James Hillman would call "archetypal," although his purpose is quite different from theirs. At a recent conference at the Dallas Institute, celebrating the life and works of James Hillman, the inheritor and uniquely original interpreter of Jung's theories, I was crucially aware of Bachelard's continuing influence. Hillman was devotee of Bachelard's exploration of the image from the inside out. He frequently quotes the French philosopher in all the volumes of his *Uniform Edition*, which the Dallas Institute coproduces with Spring Publications. The conference concentrated on Volume One, entitled *Archetypal Psychology*, and launched an annual event focused on Hillman's canon. The following quotation from this volume gives direct evidence of how important Hillman finds Bachelard's methodology, especially his intimate relationship to poetic images of the world:

> Finally, a psychoanalysis of the phenomenal world is based less on phenomenological method or on systems theory of interdependence

than on the poetics of Gaston Bachelard. There
is an elemental reverie, a mythical imagining
going on in the world's stuff much as the soul of
the human is always dreaming its myth along.
Things transcend themselves in their affordances
(Gibson), in their imaginings which poets from
Wordsworth and Coleridge through Borges,
Williams, Barthes, Ponge, Oliver, Blakeslee, and
Bly . . . make very clear. Things offer themselves
as animals do to one another in their display.
Substances themselves project upon each
other according to the alchemical definition
of projection. Not the human subject, but the
images invent the ideas we "have." They come in
(*invenire*) . . . [16]

This sentence seems crucial: "There is an elemental
reverie, a mythical imagining going on in the world's stuff
much as the soul of the human is always dreaming myth
along"; it provided the title for this book. I am including
the following excerpt from the presentation I gave at the
conference because it emphasizes that Bachelard's ideas are
alive and still reverberating:

Traditionally, science and poetry are cast as opposites.
Jerome Bruner argues tellingly that: "The scientist and
the poet do not live at antipodes. The artificial separation
of the two modes of knowing cripples the contemporary
intellectual as an effective myth maker for his times."[17] And,
indeed, that is one of the unique qualities in Bachelard
himself: no separation of science and poetry. The famous
mid-twentieth century Sorbonne professor, philosopher of
science, writer of multiple scientific volumes, can also be so

thoughtful on the subject of creative imagination.

Bachelard, like Hillman, deplores the lost vitality entailed in the current attitude toward the world around us, the lack of awareness of so many of the textures and colors and characteristics of the intimate world that surrounds us, of the tensile dimension of human nature. In *Earth and Reveries of Will*, Bachelard argues for a shift in attention: "To put the lost dream back into words, we must return in all innocence, to things."[18] The earth is the best teacher we have.

Bachelard grasped how as a condition of regarding the world in cool, scientific ways exclusively, and what followed—the breaking of direct connections with the natural world and resulting abstractions, with the artificial remove from the thingness of things—we have lost much of the sensate satisfaction of the material world. In his words: "[Things] appear indifferent to us simply because we regard them with an indifferent eye. But to the bright eye everything is a mirror; the sincere and the serious look sees depth in all things."[19] Hillman, too, urges using imaginatively precise words—"thing words, image words, craft words."[20]

All is not lost, however, as we have our painters and poets who gaze deeply and provide us with startling images. Hillman further remarks: "[L]ove itself is in images, their creative appearance and their love for that particular human soul in which they manifest"[21] or, to borrow the same thought from a line of the poet Richard Wilbur: "Love calls us to the things of this world."

Bachelard repeatedly expresses his wonder, even love, for the image, especially the poetic image. An image in its sudden appearance, particularly those that "appear to be a concentration of the entire psyche," can grab our attention, but Bachelard's acute observation is that the most effective ones do not follow just any random pattern. There is an imaginary logic to images.

In his *The Poetics of Space* and elsewhere, Bachelard always insists that poetic images emerge from an intuitive awareness of their patterns of alikeness and contrast. The essence of the poetic image is not from some subliminal source, but follows its own destined logic: "Poetry comes naturally from daydream, which is less *insistent* than a night-dream; it is only a matter of 'instant's freezing.' . . . The archaeology of images is thus illumined by the poet's swift, instantaneous image. . . . [I]mages are incapable of repose."[22]

For me, this quality explains why a brilliant poetic image is such a splendid example of human imagination at work: it is swift, dynamic, hitting home in its flashing impact. It garners multiple associations. Bachelard was in advance of his mid-twentieth century time. In detailing the impact of a singular image, he discarded Aristotelian logic: A+B delivers C. Instead, he insisted that an image works by its power to enlarge, to expand with a variety of associations. It is like a drop of water spreading out and enveloping in ever-larger concentric circles. This dynamic movement stimulates pleasure by virtue of its expansion, as it engages the reader's responding imagination. It is in the very nature of imagination to stimulate expansion, and even excess. Bachelard's words: "That is why, despite their excesses, such images are most apt to reveal imagination's latent powers By following the traces of such images, we may be able to identify the moment they begin to taper off into metaphors."[23]

Bachelard disapproves of the psychological tendency to reduce an image down to a simple metaphor. "There is more to reading an image than lies in strip-mining it down to its psychic origins," according to François Pire, one of Bachelard's contemporary critics.[24] To enter into the image is what Hillman likes that Bachelard does. "Resonating" with the poet's imagination is his constant urging.

I could go on and on about love and the image, but will conclude in reminding you finally in Hillman's words that: "The soul is always wanting . . . and this is the reason for the necessity of unfulfillment and why a psychoanalysis of anything to do with soul is, as Freud said, interminable."[25]

Bachelard demonstrates, by many examples, why our imagination is gathered in the reverberatio‚ns of elemental images. Wherever we find multivalence—for example, lively attraction and its opposite, repulsion, or loving and fearing simultaneously—we are in the sway of a dynamic image. In *Fragments of a Poetics of Fire* Bachelard reminds us, "The more brilliant an image the more troubling its ambiguity, for its ambiguity is that of the depths."

As a byproduct of his extensive range, Bachelard's methodology tutors us in how to exist circularly, being alert to all associations, with an enlarged inclusiveness, and thus, surprisingly, to live more blissfully. Just reading him makes me feel glad and thankful for the materiality of the world close at hand. Bringing imagination, which by its nature is expansive, to any endeavor relieves the despairing, dead-end sense and stress of narrow, confined possibilities.

Finally, it is his vitality, his freshness, his sometimes radical use of language, and his joyful, uplifting voice that goes along with his love of the texture of the actual, material world. His conviction that matter matters awakens a sensuous restoration of physicality to the rationalistic view of the earth as an unliving planet. At the same time, his kindness reveals an attitude that is deeply humanizing. Like returning to a favorite poem over and over again, it is possible to read and re-read Bachelard, always finding something novel, always finding spiritual renewal.

CHAPTER ONE

CALLING US TO THE THINGS OF THIS WORLD

Things answer our gaze. They appear indifferent to us simply because we regard them with an indifferent eye.

With awe and wonder Gaston Bachelard approaches his explorations of the physical world. For him the objective world is more than an inert scientific sphere. It is alive and responsive. It challenges the human being to participate. Through interaction with the world we learn about our soul's desires. Nature mirrors our spiritual aspirations. The world is literally Bachelard's cathedral. He probes the multiple manifestations, magnificent or frightening, of commonplace matter. Like a verbal alchemist unveiling the components of the material world, he teaches that spirit is the heart of matter; that matter is the ensoulment of spirit. For Bachelard, the world is constantly enticing human beings into an active relationship, one of contact or even contest, resulting in a challenge to our highest skills and strengths.

Bachelard is not an easy read. He talks in circles and

sometimes seems to be going nowhere. The experience is more akin to puzzling out a poem. Though actually in his lifetime a distinguished philosopher of science at the Sorbonne in Paris, with many scientific books to his credit, he is also a philosopher/psychologist, an explorer of levels of consciousness detailing multiple complexities of what it means to be a human being. He is a pioneer in language arts as well, a discoverer of novel nuances of language and thought. He accomplishes this feat by concentrating on a few specific images and allowing gathered associations to accumulate. He is maddening in that his words can't be read quickly or casually scanned. To follow all his meanderings takes the kind of concentration that a poem requires. And yet his interests are always connected to matter, to what he calls in *Flame of a Candle* the "psychology of the familiar, the rapprochement with real objects, the *friendship for things.*"

Bachelard charms us by drawing us nearer, closer to the material world. We may have thought before that we knew our natural surroundings quite well, but I maintain that we were wrong in this assumption. In essence, we never knew how specifically the elements of earth, water, air, or fire could evoke such attraction until we follow along with Bachelard's musing and allow our imaginations to carry us into the enticing connections he points out. As when we actively absorb a poem or a painting, when a poet or a painter focuses our attention and provides an opening to whole new vistas of participation for us, so too Bachelard moves us out of our narrow, myopic confinement into expansive engagement with the world. For example, like the poet Gerard Manley Hopkins, we are struck by the gliding flight pattern of a small bird, "stirred by a bird /

the majesty of the thing"—the etching motion of a bird in flight against the sky, in "The Windhover." Here the arching path of the bird brightens the observer's glance. With more sensitivity, more acute observation of detail, we are gathered into the enchantment of the world around us.

Responding and engaging, our imagination soars and our mood elevates. In effect, our heart gladdens at the sight of beauty: "While images of form and color may well be sensations that are transformed, material images engage us in a deeper affectivity and this is why they take root in the deepest layers of the unconscious." This constellation of object and subject connects us intimately with our natural surroundings: "Material images substantialize an *interest*,"[1] Bachelard declares. The subject, or viewer, and the object—a bird, in Hopkins' case—affect each other.

In extolling its inspiring quality, Bachelard provides the explanation of how imagination and the living world are coupled together. "The imagination is nothing other than the subject transported into things. Images bear therefore the mark of the subject . . . . And this is so clear a mark that in the end it is by means of images that the most accurate diagnosis of the temperaments can be made."[2] So the images we carry bear our own special stamp. After we get used to this idea of the interconnectedness of the viewer and the viewed, one of the most surprising aspects of this statement is the predictive aspect of material images. Bachelard claims that, though we may respond to many different elements in nature, we can learn much about ourselves by observing those that we especially identify with. Besides *Earth and Reveries of Will* and *Earth and Reveries of Repose*, he wrote many books clarifying the connections with the other

3

three elements—air, water, and fire. He again verifies this point of view: "An image that ranks as a fundamental image becomes the fundamental matter of our imagination. This is true for each of the four elements."[3]

Bachelard frequently points out the copious use of a specific elemental image by an individual poet, citing Nietzsche as a poet of high, cold air in *Air and Dreams*: "For Nietzsche, in fact, air is the very substance of our freedom, the substance of superhuman joy. Air is a kind of matter that has been mastered, just as Nietzschean joy is human joy that has been mastered . . . . *Igneous* joy is love and desire—*aerial* joy is freedom."[4]

Let me emphasize this: igneous joy is love and desire. Bachelard doesn't use W. B. Yeats as an illustration, although it strikes me how often Yeats uses fire imagery. These lines are part of the third stanza of his famous poem "Sailing to Byzantium": "O sages standing in God's holy fire / As in the gold mosaic of a wall, / Come from the holy fire, perne in a gyre, / and be the singing-masters of my soul." Also, in the companion poem "Byzantium," the fourth stanza intensifies the use of fire imagery as a way of refining the complexities of the human spirit:

> At midnight on the Emperor's payment flit
> Flames that no faggot feeds, nor steel has lit,
> No storm disturbs, flames begotten of flame,
> Where blood-begotten spirits come
> And all complexities of fury leave,
> Dying into a dance,
> An agony of trance,
> An agony of flame that cannot singe a sleeve.[5]

In connection with the image of fire, it is interesting to note Bachelard's reference to C.G. Jung and his studies in alchemy: "If we read Jung's lengthy study of alchemy,[6] we shall reach a fuller understanding of the dream of the depth of substances. Indeed, Jung has shown that alchemists *project* their own unconscious onto the substances they have long been working on and this unconscious then accompanies sensory knowledge. When alchemists talk about mercury, they are thinking 'externally' of quicksilver while at the same time believing themselves to be in the presence of a spirit that is hidden or imprisoned in matter."[7] One of Yeats's most fetching poems, "The Song of Wandering Aengus," exemplifies the alchemist's search for spirit in matter. Hunting in the woods because of the "fire" in his mind, the alchemist hooks a berry to a thread and catches a "little silver trout." After he lays it on the floor, he turns away to tend his fire, but then hears something rustling and calling his name:

> It had become a glimmering girl
> With apple blossom in her hair
> Who called me by my name and ran
> And faded through the brightening air.[8]

Though he is old with wandering, he will find out where she has gone: "And pluck till time and times are done / The silver apples of the moon, / The golden apples of the sun." The use of fire imagery, with the metamorphosis of the silver trout into a glimmering girl, results in an unending search for the silver and gold within the natural world and exemplifies the alchemist's use of quicksilver to unveil the spirit of gold in the world of matter.

Bachelard mentions Teilhard de Chardin, the metaphysical theologian, in *Flame of a Candle,* claiming a kinship with him. Like Bachelard, de Chardin, to borrow the words of Christian de Quincey, "spoke of the 'within' of things, a psychic, subjective complement to the external forms and energy of atoms, cells, plants, and animals."[9] In this passage by de Chardin, in rhapsodizing about matter, the image of fire and flame is unmistakable:

> Throughout my whole life, during every moment I have lived, the world has gradually been taking on light and fire for me, until it has come to envelop me in one mass of luminosity, glowing from within . . . the purple flush of matter imperceptibly fading into the gold of spirit, to be lost finally in the incandescence of a personal universe . . . . This is what I learnt from my contact with the earth—the diaphany of the divine at the heart of a glowing universe, the divine radiating from the depths of matter a-flame.[10]

In *The Right to Dream,* a book of essays on the arts, Bachelard says that we, like poets and painters, are "solicited by the elements." To read his books on the elements is not to acquire more knowledge, but to change one's way of looking at the material world. Bachelard approaches the world with an attitude of awe and wonder. He values direct experience of the material world because the world beckons us to participate. It is hard to decide which response he prefers—the earth that beckons us with a call to action, or the after-reflection that provides endless metaphors in which the role of imagination is more obviously entailed. He is an acute observer, one who discovered much to celebrate in close involvement with the textures, fabrics, colors, and movements of the tangible world. He puts forth this unique opinion in *The Right to Dream*: "Things answer our gaze. They appear indifferent to us simply because we regard them with an indifferent eye."

Bachelard could be accused of projecting human emotions upon an inert world, but he argues unconditionally for just this point of view. To anthropomorphize the world is to discover that everything in the world is alive: "The meadow is not a mantle of greenery, it is the earth's primordial will."[11] Like Dylan Thomas's poem "The Force That Through the Green Fuse Drives the Flower," Bachelard recognizes the joining of the life force, the livingness of the flower, and the consciousness of human being. Calling us to the things of this world, he never lets us forget the interconnectedness of all.

CHAPTER TWO

WHY MATTER MATTERS

Matter reveals to us our own strengths, suggesting a system for categorizing our energies.

Bachelard's singular contribution, what is unique about him, is the way he gives renewed meaning to the material world. It is a "felt change of consciousness" (Owen Barfield's term), a sudden awareness that the natural world is full of joy if we but look—and always, in some ways, unpredictable and a mystery. In touch with matter, we experience a sudden lifting of any sadness of the heart. I think of it as a quickening, an *in-spiriting* of matter. This proclamation of renewal is similar in sentiment to Robert Frost's poem "Kitty Hawk": "Spirit enters flesh / And for all it's worth / Charges into earth / In birth after birth / Ever fresh and fresh." Or closer still, like the celebration of Gerard Manley Hopkins's "God's Grandeur" when he addresses the "dearest freshness deep down in things." Perhaps it was Bachelard's peasant background, which he never tried to shed, or his belief that the scientific approach disregards the livingness of matter, that carried his interest deeper into the materiality of the things of this world. Even in the field of the philosophy of science, his writings tend away from the totally abstract toward an engagement with the world. This desire to connect the things of the world with the profound depths of what it means to be human explains Bachelard's dual, sentient interests—in poetry and in science.

For Bachelard, the earth is a classroom where we learn how nature functions, but also the *in situ* occasions where we earn our identity. The words "human nature" directly link us to the natural world. It is in our interaction with the world that we find the bonds that strengthen our resolve to survive. Ever since Hesiod, who gave Mother Earth the primal role of origination—even the birthing of the heavens—the ancient Greek world felt its identity with matter. In the *Theogony*, Hesiod describes the beginnings of the universe: "Earth, the beautiful, rose up / Broad-bosomed, she that is the steadfast base / Of all things." The earth, Gaia, matter, mother—all are intimately associated with our personal beginnings. It is only later that we separate ourselves from the materiality of our environment, or think that we can. The earth gives us a sense of who we are and locates us in the cosmos. In the words of Rainer Maria Rilke, it is in the earth where we find the transformation of even bitter experience into valuable meanings of identity and destiny: "To the still earth say: I run / To the swift water speak: I am." Bachelard echoes this sentiment in *Earth and Reveries of Will* in saying that "Matter reveals to us our own strengths, suggesting a system for categorizing our energies." Matter is our testing ground, or, in a lyrical moment, as Frost says: "Earth's the right place for love: / I don't know where it's likely to go better."

It is basically in its role as teacher that Bachelard exalts terrestrial matter. Again in *Earth and Reveries of Will*,

9

Bachelard makes one of his typical inversions: "Matter is the mirror of our energies; a mirror that focuses our strengths by illuminating them with imaginary gratifications." The world of matter evokes image-pictures that energize us to make gargantuan efforts. The earth gives us a sense of up and down, inside and out, above and below. According to Bachelard, gravity gives us the longing for lightness. Earth presents challenges to us, enticing our interaction with, or, more dynamically and actively, provoking reactions to an array of possibilities for engaging our attention. This active, creative potential is inherent in the earth's dynamism, not simply waiting to be revealed by human action. Earth is an inexhaustible mine of riches to be discovered. Presented with sand and water, we are inspired to make a clay vessel. Finding a river, we are compelled to build a bridge. We are not allowed to be passive in our engagement with matter. That is the crux of it—matter engages our imagination and summons the personhood of each of us, evoking the will to action. Imagination, like an ever-mobile radarscope, expands and takes in all that surrounds it, being supercharged by contact with the dynamic material world. "Even in contact with an element as strongly material as the earth, imagination asserts its mobility," Colette Goudin explains in her introduction to Bachelard in *On Poetic Imagination and Reverie*.[1]

The title of the first Earth book, *Earth and Reveries of Will*, brackets together one of Bachelard's most startling oppositions—dreams or reveries on one hand, and the will-to-action on the other—seemingly contradictory terms. He designates these unlikely combinations as "reveries of will." Generally the acts of willing and dreaming are thought of as opposite urges, with dreaming considered passive and ineffective compared to the energy of action. Bachelard argues that matter tests us and schools the will: "It [matter] provides not only enduring substance to our will, but a system of well-defined temporalities to our patience. In our dreams of matter, we envision an entire future of work; we seek to conquer it through labor. We take pleasure in the *projected* efficacy of our will."[2] Bachelard makes a good case that there would be no effective will without the impetus of the dream behind it, because "to dream material images— yes, simply to dream them—is to invigorate the will. It is impossible to remain distracted, indifferent, or detached while one dreams of resistant matter clearly imagined."[3]

Bachelard reminds us that matter is not always easily malleable to our wishes. To the contrary, "Matter is the best measure of our hostility." A hard surface or any recalcitrant matter thwarts our efforts and only yields when will is insistent. We have no idea, until we actually try it ourselves, how resistive matter really is. Bachelard gives this amusing example of fashioning an article in wood: "One wishes to file in a straight line, *imposing* rectilinearity. But matter, for its part, has its own desire to retain a certain roundness in mass and line. It refuses, with an undisguised *bad will*, to yield to elementary geometry. Only the worker knows by what adroit tactics and what calculated restraint one manages to impose simple form on an object."[4]

With the advent of psychology, we have been led to believe that our emotions are honed primarily through our personal relationships, especially with parental figures and siblings. In *Water and Dreams*, published in 1942, Bachelard was already saying that will arises in response to the material world as well as in the human encounter: "Man is an integrated being. He has the same will against all adversaries. All resistance awakens the same wish. In the realm of will, there is no distinction to be made between things and man."[5]

Bachelard presents the twin poles of the psychology of work as "the intellect and the will—clear thought and dreams of power. It must first be understood, however, that when it comes to working hard matter, the actions of these two poles are inseparable."[6] In both work and play, matter sorts out our determination. In these activities we become aware of the value of patience, but equally important, the beneficial use of our anger, which Bachelard considers an invigorating asset when applied to a conquering endeavor. He quantifies the anger required for different types of creative endeavors with the engraver requiring the most fervent intensity to inscribe the plate:

> The engraver is in a state of perpetual rebellion against restraint. Through his every joy there runs a streak of anger. Before, during, and after his work the eyes, fingers, and heart of the good engraver are pressured by rage. The burin needs this hostility behind it, this darting fury, this trenchancy, this incisiveness—this decisiveness. Once again, every engraving bears witness to a certain force. Every engraving is a reverie of the will, an eager outburst of the will to construct.[7]

In his exploration of each element, Bachelard keeps us grounded by including the unseemly, dark, or unseen side, well aware that we dream not only in technicolor, but also in sepia. Following this practice, he reminds us: "In the dreams of the terrestrial psyche, the earth is, in effect, dark and somber, grey and lusterless: the earth is earthen." In summing up the totality of the primordial image of a flower, blossoming is linked with dung, death, and the earth:

> The flower is without doubt a primary image, but for one who has worked the soil, this image is dynamic. Participation in the mysterious workings of dark earth helps us better to understand the gardener's reveries of will that associate the act of flowering with the act of embalming, producing the brightness of lilies along with the darkness of mud.[8]

Bachelard's observations fortify the bittersweet sense we always have in the presence of a gorgeous flower—the awareness of its brief beauty. The earth's creativity, which he extols, is matched by an equal thrust—destruction—required for renewal. Mother Earth who births us is also our burying ground. George Meredith echoes the closeness of these contrary dual aspects—creation and disintegration: "Earth knows no desolation / She smells regeneration / In the moist breath of decay."

It is not surprising that, in gathering together central images, Bachelard magnifies the entire spectrum of qualities from positive to negative. "Images attract and repel one another with a sublime reciprocity that is the very life of the imagination."[9] He insists that the mobility of imagination—one of its principal characteristics—depends upon the flow of energy between contrasting opposites, for "no association can exist without its opposite." Images, more stirring when they gather together associations of active ambivalence, are more energizing than ideas: "Ambivalence in images enjoys a far more active function than antithesis in the realm of ideas." This point of view is significant personally for me. With a sun sign in the zodiac of Libra, or balance, Librans are always accused of seeing both sides of any argument, of weighing every point of view before making decisions. Perhaps being pulled back and forth is appropriate and to be expected whenever imagination is engaged. For Bachelard, that is the rule: "Imagination, ever provocative, engages in combat." He always questions established thought, challenging us to take nothing for granted. He believes that error is efficacious and that we only arrive at truth through confrontation. Mary McAllester Jones explains: "Bachelard was always a polemical thinker, believing, as he declared in *The Philosophy of No* (1940), that 'two people must first contradict each other if they really wish to understand each other. Truth is the child of argument, not of affinity.'"[10]

Although all uses of the imagination excite him, Bachelard marvels most at the creative energy of the poet who, in the written word, can yoke together ambivalent tensions. Literary images are revealing for him: "I love man, most of all, for what can be written about him. Is what cannot be written about him worth living? I have to be content with the study of material imagination that is *grafted* on. I have nearly always limited myself to studying branches of materializing imagination *above the graft* after culture has put its mark on nature."[11]

Bachelard bestows loving, lingering concentration

on the dynamic poetic image, allowing it to reveal itself leisurely with all its reverberations. As he succinctly puts it in *The Poetics of Reverie*, focused concentration is demanded, and this requirement is one of its major attractions: "One doesn't read poetry while thinking of other things." Slowing us down, the way reading poetry slows us down, makes it difficult to ever skim a few pages in Bachelard's works. Through such total attention, the reader communicates with the poet. Bachelard's phenomenological method, entering wholeheartedly into the image, "leads us to attempt communication with the creating consciousness of the poet. The new-born poetic image—a simple image!—thus becomes quite simply an absolute origin, an origin of consciousness."[12] In other words, through empathetic engagement with the birth of a well-chosen image, we feel renewed and more presently alive.

It is Bachelard's "lovely concreteness" (Robert Sardello's term), at once simple and complex, that is riveting. Eagerly he grabs experiences that might seem mundane and endows them with the shine of the imagination. Here is one of his delightful recommendations:

> When insomnia, which is the philosopher's ailment, is increased through irritation caused by city noises; or when, late at night, the hum of automobiles and trucks rumbling through the Place Maubert causes me to curse my city-dweller's fate, I can recover my calm by living the metaphors of the ocean. We all know that the big city is a clamorous sea, and it has been said countless times that, in the heart of the night in Paris, one hears the

ceaseless murmur of flood and tide.[13]

Bachelard urges us to allow the world to engage us—its textures, its colors, its forms. He often amazes with his accuracy of observation. In the following exchange, he reveals his sense of humor about the quality of coldness:

> The imagination of cold turns out to be rather poor . . . . Why such poverty? It is probably because nocturnal life lacks a true oneiric sense of cold. It is as if humans asleep are entirely unconscious of low temperatures. When on occasion it happens that I am served wine in my dreams, not only does it always turn out to be tasteless, even more horrifying for a man from Champagne it is always brought to the table *at room temperature*. Quite naturally it is as warm as milk! In dream it is impossible to get anything cold to drink."[14]

Bachelard, without preaching but simply by sharing his own joy, awakens the simple pleasures in the materiality of the physical world. Were we to take his sensibility to heart, it would become anathema to mutilate our precious planet. His message is vital for the twenty-first century, which is so often myopic and forgetful of the interrelatedness of all in the cosmos. In the 13th century, the great mystic Hildegard of Bingen spoke of the links between all levels of being: "Everything that is in heaven, on the earth, and under the earth is penetrated with connectedness, penetrated with relatedness."[15] Three centuries of mechanistic science have dulled our awareness of our intimate relationship with the environment. Rhapsodizing on the excitement of experiencing the everyday pleasures surrounding us—in sky, in rocks, in dew, in streams—Bachelard calls us to the "things of this world." With his love of the soil of the earth, Bachelard reminds us to value every sod.

By arousing respect for matter, Bachelard provides a new stance, more compelling than intellectual, a deep desire to care for the universe, a truly active sense of the word "ecology." I think of him as an early environmentalist. I take the stance that we desperately need the imaginal ecology that Bachelard provides. The imagination of matter, which he so attentively supplies, comes first, before we can genuinely seal a pact of care with our environment. Only by living in the lively imagination of matter does the imagination work deeply into our feelings and transform us. Experiences such as taking in a lovely sunset, while pleasurable, can be only fleeting impressions. Outer ecology, such as deploring the manmade debris in space, while important, can remain in the realm of ideas, sometimes quite abstract. But images—charged with Bachelardian inflection—change us.

CHAPTER THREE

THE SEDUCTION OF MATTER

Whenever we are struck by an image, we should ask ourselves what torrent of words this image unleashes within us.

Bachelard, a philosopher of science who in 1938 became interested in how the imagination works, continued exploring his dual interests until he died in 1962. The date 1938 is notable because this was the year he published his first book on the analysis of an element, *The Psychoanalysis of Fire*. Attracted to the early writings of Freud, he attempted to bring the psychoanalytic approach to the understanding of a physical property. This book marked him as a most original and contrarian thinker. In order to be accepted by a thinking world that trusted "facts" before all else, psychology at the time was trying very hard to hold that it was part of the scientific field and not a part of the humanities. Bachelard, on the other hand, insisted that while science and poetry were distinct and sometimes opposed ways of thought, these differing ways of approaching the same reality were united in the use of the imagination. Bachelard had reached the amazing conclusion that the imagination is a fundamental human characteristic and underlies all ways of viewing the physical world. In other words, the scientist and the humanist each see reality filtered through an imaginative, refractive discipline. However, he constantly reminded his students and readers that the essence of the imagination is its mobility, which means that to reduce any category of thought to a singularity is a gross misunderstanding. Later in this chapter, we will return to the idea of active imagination—or reverie—and dream, and what is really real for Bachelard.

After *The Psychoanalysis of Fire*, Bachelard would continue writing about science, but nine books later, and until the day he died in 1962, he was equally fascinated by images and the imagination. His unfinished manuscript, *Fragments of a Poetics of Fire*, which the Dallas Institute has translated, with explanations by his daughter Suzanne, testifies that the subject was not exhausted, that he had not been able to say everything about the element of fire, even though in 1961 he had published *The Flame of a Candle*. He would not agree that the imagination is a mere idle pastime, nor would he agree with those who claim to have no imagination. For Bachelard, the imagination is fundamental, underlying and overlaying *all* perception. The imagination is not of a random nature but has patterns of organization that can be followed. These configurations are what Bachelard explores. He claims that "The imagination is more *determined* than we think; even the most contrived images are governed by rules. In many respects, the theory that human imagination is structured by the four elements returns to a study of *determinism of the imagination*."[1] Throughout this chapter I will be following Bachelard's manner of weaving commentary on an individual element with a continual return to the consideration of the very nature of images and of the imagination, which he considers so vital.

Following his first foray into the nature of fire, Bachelard continued the same detailed study of each of the imaginative associations of the three other classical elements—water, air, and earth. Though often arguing with Edmund Husserl, the father of philosophical if not psychological phenomenology, Bachelard applies some of the principles of phenomenology—examining each phenomenon with fresh eyes—with the realization that consciousness is always "intentional," always consciousness of *something*. Bachelard's powers of observation are as keen as those of a painter or poet. He unceasingly calls attention to the texture and subtleties of the material world that, because we see it every day, we so often take for granted. We may think this acuity is a simple matter of opening up our awareness, but I feel a lingering sense of gratitude for being reminded by him in *Earth and Reveries of Will* of the simple joys of active participation in the physicality of the things that surround us, for "material reveries root us in the universe." Or, as he elaborates: "Dreaming before the fire or water, one knows a sort of stable reverie. Fire and water have a power of oneiric integration. Then the images have roots. In following them, we adhere to the world; we take root in the world."[2]

Bachelard's study of the elements provides not only immediate pleasure, the "Ah ha!" of a sudden recognition of renewed connection with the material world, but also a way of classifying different personalities. Each element is like a point on the compass, or like a direction on the

Native American medicine wheel, revealing where you are. In *Water and Dreams* he calls the four fundamental elements "signs of *philosophic disposition*." So let us explore first the two "feminine" elements—water and earth—and then the two "masculine" ones—air and fire. And just to complicate matters, Bachelard considers water an in-between element, in between "fire that puts it out and earth that needs it to sprout."

In 1942, when he published *Water and Dreams,* Bachelard was getting prepared for an extended comparison of the four material elements. He begins the first chapter with this statement: "'Images' whose basis or matter is water do not have the same durability and solidity as those yielded by earth, crystals, by metals, and precious stones. They do not have the vigorous life of fire images."[3] In the introduction Bachelard announced his plan to include a new theory of literary imagination based on the study of the *pure literary image*. On this point, he faced immediate objections. He was accused by his detractors of snatching images from poetry, while ignoring the entire poem, a criticism that was certainly true. However, explication of whole poems or whole works of a poet was not his goal. He wanted to find a way to account for the power of radiation of a single poetic image. The creativity of a poet was transporting for him and could even be elevating enough to relieve unhappiness. Here is his sentiment that I always recall when I am trying to dispel any distress: ""For anguish is factitious: we are made to breathe easy. And it is in that way that poetry—summit of all aesthetic joy—is beneficial."[4]

In several following books Bachelard expressed his desire to renew literary criticism (which certainly was laudatory, the field having strayed so far away from the materiality of the text). "If my research is to have any impact, it should contribute some means, some tools, for renewing literary criticism."[5] His main thesis was that we should *participate* in literature, not merely study or analyze it. He himself seized literature, advocating a second or a third reading of what one loves. He believed that the pleasure of reading was being fully caught up in a transporting experience. In his way of thinking, author and reader are engaged, and even co-creators, in a jointly shared understanding. The well-chosen literary image illumines the world we take for granted. He thrills to those images "wholly brought to life in literary texts, or at least to images that would remain inert were they not given more elaborate expression. Often, when we consider simple dull reality, it is entirely lacking in those attributes with which it is *transfigured* in literary imagery."[6] He mentions as an example: "The poor lark is unseen in flight, its song monotonous. But as numerous literatures attest, it is a sort of cosmic center so exhaled as to provide the symbol for a state of the soul and a sunny universe at the same time."[7]

Bachelard renewed his warnings against "the temptation to study images as *things*. Images are 'lived,' 'experienced,' 're-imagined' in an act of consciousness which restores their timelessness and their newness."[8] Colette Gaudin remarks on his method of participating with the poet in reliving a poetic image. Repeatedly, Bachelard questions this surprising paradox: "How can an image, at times very unusual, appear to be a concentration of the entire psyche?"[9]

First comes the arousal of attention—an image that ignites our interest; then, an expansion in random directions of its imaginary aura. Fundamental to Bachelard's method is the technique called "active imagination" in Jungian psychology—allowing oneself to be carried along with the flow of associations connected with any thought or image. In the introduction to *Air and Dreams*—and one could learn much about Bachelard just by reading his introductions—he explains:

> Whenever we are struck by an image, we should ask ourselves what torrent of words this image unleashes within us. How can we detach it from the all too stable background of our familiar memories? To grasp the imagining role of language, we must patiently search out for every word its inclinations toward ambiguity, double meanings, metaphors. To put it in more general terms, we must take account of every urge to abandon what we see or what we say in favor of what we imagine. In this way we may be able to reinvest the imagination with its role as seducer. Imagination allows us to leave the ordinary course of things. Perceiving and imagining are as antithetical as presence and absence. To imagine is to absent oneself, to launch out toward a new life.[10]

Water

Bachelard claimed water as his own favorite element. He called it "feminine," because of all the elements, it is the most receptive to various agents acting upon it. For instance, wind stirs up the turbulence of the stormy sea, and earth can muddy or pollute pure water. He explains the feminine connection: "We shall also see how profoundly *maternal* the waters are. Water swells seeds and causes springs to gush forth. Water is a substance that we see everywhere springing up and increasing. The spring is an irresistible birth, a *continuous* birth."[11] Then, he adds that "the unconscious that loves such great images is forever marked by them." But Bachelard also cautions against any attempt to narrowly confine a vital material image. Although his way of describing a primal image shares some similarity to C.G. Jung's definition of an archetypal image, there are some major differences in that Bachelard did not emphasize the historical or mythological dimension to the same extent that Jung did. I might add that, though Bachelard started out with Freud, he felt much more kinship with Jung. He referred often to Jung in the later books and said that he regretted not knowing more about Jung earlier.[12] However, he always took psychology to task for not exploring fully the dynamics of either will or imagination.

In *Water and Dreams*, as Edith Farrell discerningly points out:

> Truth became less restricted, and Bachelard began to work positively with literature, hoping to find bases for an objective study of literary imagery. More and more, as he wrote on this and the other elements, poetic truth acquired a life of its own.
>
> No longer was an image to be analyzed, but a joy with which to participate. No longer was Bachelard content to psychoanalyze; he began to find the essential liberty of the image

which escaped classification and the liberty of the dreamer to follow his dreams. In these last books, based on the phenomenological method, this liberty became the most important factor in poetic imagination. The image was allowed, was encouraged to escape from any attempt to fit it into a scheme. Bachelard then lived the images in the long, slow reverie in which he found so much enjoyment and the value of which he communicated to others.[13]

Bachelard's affection for poetry and the well-chosen poetic image is manifest in every book. The poet brings to expression our innermost thoughts, or, put another way in *The Poetics of Reverie*, "Poetry nourishes within us reveries which we have not been able to express." The poet does not describe, he exalts things. However, Bachelard lays to rest any notion that valorizing the world, poeticizing it, is merely a passive response to a pleasant presentation. He assures us that the poetic act is essentially another manifestation of the will-to-power: "The conqueror and the poet both want to put the brand of their power on the universe. They both take the branding iron in hand; they brand the dominated universe."[14]

The element of water provides a perfect example of the double-sidedness of any vital image. Although water is predominately "feminine," this is in no way an absolute, as water can become violent. Bachelard, like Jung, insists that all images truly active in the imagination carry double valences—active and passive, masculine and feminine, light and dark, with both reassuring and frightening aspects.

Without equivocation, he claims in *Water and Dreams* that is a "primordial law of the imagination: *a matter to which the imagination cannot give a dual existence cannot play this psychological role of fundamental matter.*" There must always be a "*dual participation* of desire and fear, a participation of good and evil, a peaceful participation of black and white." Consciousness proceeds by discrimination between opposites, but any image that truly resonates returns us to the duality. Bachelard provides a pathway and encourages us to always remain with the tensions of opposites. It is stultifying to the imagination to be bound to only the black or white end of the spectrum. Our participation in the imagination of these great material images schools us in the subtleties and complexities of things around us, allowing us to enjoy them more deeply. In *Water and Dreams*, he notes: "Material imagination learns from fundamental substances; profound and lasting ambivalences are bound up in them."

Water is the very substance of existence. Our bodies are more watery than solid. Throughout *Water and Dreams*, Bachelard won't let us forget that, though life arises out of water, death at the end of life returns us to the watery deep: "Water mixes its ambivalent images of birth and death" and therefore "water is the *melancholizing element.*" The element of water is more closely associated with final demise than any of the other three: "Each of the elements has it own type of dissolution, earth into dust, fire into smoke. Water dissolves more completely. It helps us to die completely."[15] Water is both source of life and, in our lively imaginations, also the place of return. He explains: "Water, the substance of life, is also the substance of death for ambivalent reverie." We speak of the breast of the wave, which somehow doubles

26

the maternal image of the sea.

In death by water, Bachelard quotes Jung when explaining that "the dead person is given back to his mother to be born again." He follows with this proposal: "Death in water will be, for this reverie, the most maternal of deaths. The desire of man, says Jung in another place, 'is that the somber waters of death may become the waters of life, that death and its cold embrace may be the maternal bosom, just as the sea, which, although it swallows up the sun, gives it new birth in its depths . . . . Life has never been able to believe in Death!'"[16]

In *Water and Dreams* he speaks of culture complexes that are more a part of the cultivated personality than the usual definition of a complex in psychology.

> I have given this name *(culture complex)* to *pre-reflective attitudes* that govern the very process of reflection. In the realm of the imagination, these are, for example, favorite images thought to be derived from things seen in the world around us but that are nothing but *projections* of a hidden soul. Culture complexes are cultivated by someone who thinks he is acquiring culture objectively. The realist, then, chooses *his* reality in reality; the historian chooses *his* history in history. The poet arranges his impressions by associating them with a tradition. Used well, the culture complex gives life and youth to a tradition. Used badly, the culture complex is the bookish habit of an unimaginative writer.[17]

Bachelard reminds us of the myth of Atlas, who is said

27

to have held the world up on his mighty shoulders. The hero Hercules, in one of his tests of strength, comes along and offers to share the burden, but gives it back to Atlas. To hold this story in our imaginations for a while causes us to feel stronger ourselves. Our imagination is naturally expansive. When the energy of the imagination is sensed, we begin to feel more and more powerful, and finally all-powerful. "Reveries of the will to power are reveries to be all-powerful," is how Bachelard explains it in *Earth and Reveries of Will*. That said, it follows that to imagine effort in this way makes the whole body strong, not just some muscles prevailing over others, as happens in purely physical exercise.

It was Bachelard's habit to speak of cultural complexes, for example in *Water and Dreams*, as "grafted on more profound complexes, which psychoanalysis has brought to light." Cultural complexes often relate to a figure from literature or mythology and provide a narrative structure for the better understanding of human tendencies in much the same way as parables from the Bible or stories from Greek mythology function. These complexes would only be recognized by a person with some literary background:

> I have not hesitated to designate new complexes by their cultural emblem, by the sign that all cultured men recognize; this may betray a lack of prudence, for such signs remain obscure and awaken nothing in the man who lives far from books. A man who does not read would be greatly astonished to hear the poignant charm of a dead woman adorned with flowers and drifting away, like Ophelia, with the flow of the river.[18]

THE SEDUCTION OF MATTER

In *Water and Dreams* Bachelard names two complexes connected with water and related to what he calls "the triple syntax of life, death, and water." With the image of water, Bachelard details the Ophelia complex, mentioned above and named after Shakespeare's beautiful but sad heroine's suicide in *Hamlet,* and the Charon complex, named after the classical ferryman of the boat laden with lost souls. Both complexes involve the connection of death and water, "because they both symbolize a meditation on our last voyage and our final destination." Life, with its metaphor of a journey, is always completed by a final journey over water. We find numerous examples of this last rite in literature and mythology. In Yeats's poem "Byzantium," it is a journey on a dolphin's back across "a gong-tormented sea." In classical literature, the watery River Styx surrounds the realm of the dead. All vows taken, even by the gods and goddesses themselves, on the River Styx, if broken, would result in death. In yet another instance, in the story of Psyche and Eros, when Psyche goes to the underworld she must give a coin to Charon to get to Hades' kingdom. And in Dante's *Divine Comedy*, it is again Charon who ferries the dead. Bachelard quotes Saintine who concludes: "Without Charon, hell is impossible." Other deaths are certainly always happening, but Bachelard claims that, whatever the cause, the ultimate and final journey is one over water:

> Customs based on reason may well entrust the dead to the tomb or to the pyre; the unconscious that bears the stamp of water will still dream beyond the grave, beyond the pyre, of a departure over the floods. Having passed through earth or

fire, the soul arrives at the water's edge. Profound imagination—material imagination—wants *water* to have its part in death; water is needed for death to keep its meaning of a journey. From this, we may gather that for such infinite dreams, all souls, whatever the nature of their funerals be, must board *Charon's boat.*[19]

Though Bachelard did not name a complex after Narcissus, no discussion about water would be complete without a discussion about reflection, and he spends several pages with his, as usual, original point of view about this classical story. When we use the verb "to reflect," we are using a water term. Bachelard maintains that actual water stirs the imagination more than the mirror, which we use for the more practical job of making our faces seductive:

First, we must understand the psychological advantage of using water for a mirror: water serves to make our image more *natural*, to give a little innocence and naturalness to the pride we have in our private contemplation. A mirror is too civilized, too geometrical, too easily handled object; it is too obviously a dream device ever to adapt itself to oneiric life.[20]

Bachelard broadens our awareness in *Water and Dreams*, in this case, of narcissistic tendencies: "Narcissism does not always produce neuroses." There is both an individual type of narcissism, the kind we are most familiar with, and also a cosmic form of narcissism. Both begin with the "will to appear" and engage in the dialogue between *seeing* and *being seen,* or *revealing oneself*:

Let us take the least sensual of the sensations, the visual, and see how it becomes sensual. Then let us study water in its simplest *adornments*. Following the faintest of clues, we will then grasp little by little its *will to appear*, or at least the way in which water symbolizes by means of the *will to appear* of the dreamer who contemplates it. In my opinion psychoanalytic doctrine has not given equal emphasis to *both terms* of the dialectic— *seeing and revealing oneself*—that are related to narcissism. The poetics of water will allow us to contribute to this dual investigation.[21]

Bachelard's amusing comment that the earth has its own eye (the still lake) for reflecting, takes narcissism out of the realm of an individual disorder. This cosmic eye reflects back the land. In like manner, the human eye takes delight in opening out to the material world, gazing and being seen, receptive and active, in interchange with the world. He adds:

The cosmos, then, is in some way clearly touched by narcissism. The world wants to see itself. Will, taken in its Schopenhauerian sense, creates eyes to contemplate, to feast on beauty. Is not the eye itself luminous beauty? Does it not bear the mark of pancalism? It must be beautiful in order to behold beauty. The iris of the eye must be a beautiful color for beautiful colors to enter the pupil. Without blue eyes how can one really see the blue sky? Without black ones, how look at the night? On the contrary, all beauty is ocellated. This pancalistic union of the visible and vision

> has been felt by innumerable poets; they have
> lived it without defining it. It is an elementary
> law of the imagination.[22]

We can clearly trace his pattern or methodology in *Water and Dreams*. He begins with surface images relating to water—the mirror and Narcissus—then expands with the entire universe. After, he explores reflection and idealization in the Swan complex as the image of pure, white, spotless being. Water images become more violent in the works of Edgar Allen Poe. In addition to the combination of water and death in the Ophelia and Charon complexes, Bachelard details the Swinburne and Xerxes complexes. The Swinburne complex involves the conquest of water, rather than being seduced by it. Water is a challenge. It is not a natural medium for human beings, like air, but one in which effort is required if we are to survive in it. In the action of swimming, for instance, water is subdued, a feat that reinforces the will: "More than anyone else, the swimmer can say: the world is my will; the world is my provocation. It is I who stir up the sea." We experience fear of a rough sea, for example, but are exhilarated when we dive under a wave and come out the other side. Bachelard explains: "A complex is always a hinge of ambivalence. Around a complex, joy and sorrow are always ready to exchange their eagerness."[23] Xerxes sadistically thrashed the sea for drowning his warriors, carrying dominance over water to a more intense level.

Bachelard unveils two unusual branches of the will— the will to beauty and the will to contemplate. With his repeated attention to the will to beauty and the seductive engagement of the material world, Bachelard instructs us how to see beyond our own narcissism, beyond our own image reflected back to us, beyond any projective distortions. The world unfolds as truly a place of wonder with him as our guide.

Finally, in the last chapter in *Water and Dreams,* entitled "Water's Voice," Bachelard gathers together, in his imaginative and vivid way, an element and a human capacity. In it, he addresses the subject of the sounds of language. Water flows through the words we speak, he claims: "Human language has a *liquid quality,* a flow in its overall effect, water in its consonants."[24] Like the systole and diastole of the blood, water—with its trickling and rushing, with its ebb and flow— gives us the basic rhythms of life and of language and poetry. Here Bachelard contends, "Water is the mistress of liquid language, of smooth flowing language, of continued and continuing language, of language that softens rhythm and gives a uniform substance to differing rhythms."[25] Bachelard lived in Champagne where he heard streams, and not near the sea, which explains why all of his most effective examples are of lakes and brooks. The book ends with one of the loveliest testimonies to the sounds of water I have ever read. I think of it whenever I go fishing and hear a stream descending a Montana mountainside:

> "Where is our first suffering? We have hesitated to say . . . it was born in the hours when we have hoarded within us things left unsaid. Even so, the stream will teach you to speak; in spite of the pain and the memories, it will teach you euphoria through euphuism, energy through poems. Not a

moment will pass without repeating some lovely round word that rolls over the stones."[26]

Earth

Earth is the second of the feminine images that Bachelard addresses, but the last in the series of his books. In the preface for the two volumes dedicated to earth, he includes this heading: "The Imagination in Matter and Word." This subtitle repeats his two concerns: the actual qualities of an element interacting with the individual responsive imagination, and then the expression of this material imagination in poetry and prose. In *Earth and Reveries of Will*, Bachelard explains his purpose in these studies: "I have attempted to organize and deepen the appreciation of images of fire, water, and air. Only the study of images of earth remain." Colette Gaudin's observation about the difference between images of water and earth is pertinent here: "For Bachelard each element possesses specific qualities that require particular approaches. A glance at his chapter headings is revealing in this respect. Water remains more or less similar to itself, and its diversity is expressed in adjectives: calm, clear, heavy, violent. The earth, in contrast, is viewed as different substances: rock, sand, mud."[27]

Two images that I particularly like are the sparkling dewdrop and the crystal, both capable of catching light or dreams. Each element, in addition to manifesting images able to stir us, also carries those with the ability to quiet our anxieties. A diamond, part of the crystalline reverie of earth, can give us a supreme sense of serenity. It is an image that can be said to heal: "An imagined diamond can relieve distress. Dreaming of a crystal in its matrix, it would seem, inspires one to leave feelings of unhappiness behind and to live once again in the glow of one's own light."[28]

In his preface, Bachelard explains that in order to cover the range of relationships to the earth, our home, it would require two books—one for the active or extroverted side, or the *will* at work, or force, and one for the introverted side, or *repose*. As always, both would be filled with "ambivalent certitudes." I have not been able to find out what Bachelard really intended as a subtitle for *Earth and Reveries of Will.* In one of my copies in French, there is no subtitle; in another it is *The Imagination of Matter,* but this is the same subtitle as the subtitle for *Water and Dreams.* In the list of the Bibliothèque Nationale de Paris, referring to the 1992 edition, the subtitle is *Essai sur l'imagination de la matière,* while in other places it is listed as *Essai sur l'imagination des forces,* or *Essay on the Imagination of Strength.* We are faced with the same problem with the companion volume, *Earth and Reveries of Repose.* In my first copy in French in 1977, the subtitle is clearly *Essay on the Images of Interiority or Intimacy*, which makes very good sense as the images are of the house, the grotto, the labyrinth, the serpent, the root, and the alchemist's wine. But later copies, published by the same press, José Corti, don't carry a subtitle. The French keep us guessing!

Here is Edith Farrell's synopsis of the two volumes, which she calls "the most comprehensive study of an element undertaken by Bachelard":

> In the first volume, the earth is seen as a resisting
> element, one which implies permanent hostility

to man and arouses him in his dreams of will. The opposite sides of the dialectics of hard and soft bring images of a primitive matter which man can learn to bend and shape to his will. In the dreams of work and matter, the smith stands out as an example of a heroic figure, one whose trade is "complete," that is, which involves all four elements, and to him Bachelard devotes a rather lengthy study. The second part of this first volume treats phenomena with which man is less directly involved; the rock, the crystal, and precious stones each contribute to an understanding of the mineral life which underlies all earth images. In the final section, Bachelard again takes up the images of the fall, but this time, from the viewpoint of an imagination oriented toward earth, movement toward the interior of the earth and the dynamic images of rising implies not effortless flight, as for the aerial spirit, but a struggle to maintain a heroic uprightness, whose symbol here is Atlas, a man dynamized by a mountain.

In the second volume, one finds the images which attempt to grasp the heart of matter. Not satisfied with surface images, the imagination now seeks to orient itself toward the "within" and not the "against" as was the case in dreams of will. This study concerns images of refuge, the house, the Jonas complex, the grotto, and the labyrinth, all of which show a tendency toward the psychological category of the return to mother, but which Bachelard prefers to differentiate and to which he gives a more individualized value. More dynamic reveries begin the third part, completing the images of the mineral labyrinth by the serpent and the root, all of which show similar twisting movements. The last chapter is devoted to the study of one particularly valorized image, namely wine and the vineyard. Bachelard wrote this chapter as an example of how monographs can be constructed on highly valued images.

In these volumes, the reader will note a progressive refinement of Bachelard's studies. The work as a whole, having been expanded, Bachelard is able to discuss, in more detail, the nuances of any given image. In his chapter on the root, he not only develops the primary values of archetype, but sketches the possible ramifications to which such a dream-creating word may lead. The stress once more, but even more forcefully, is on the dynamic quality of this and the other earth images. The root is a restless image. It is a serpent; it is resistance, twisted in its effort to stand firm; it entwines itself in the depths of the earth, in the realm of the dead; it sustains the tree, archetype of life; it can, therefore, never be inert. Thus, the root is "an active syntheses of life and death." The concept of synthesis which Bachelard formulates in this volume is central to his discussions. He contrasts logical and poetic syntheses, attributing to the first the value of reconciling opposites, where the synthesis is the final step of the reasoning

process and to the second that of dividing, in a
dialectic of hidden and apparent, what the image
first gave as a synthesis.[29]

It is in our relationship with earth that we are challenged
to participate with matter and create our own view of reality,
for "matter, in fact, gives us the sense of hidden depths;
it compels us to unmask superficial being." By their very
existence, material things challenge us to dialogue with
them, to probe them. The world entices us to be engaged
with it. Provocation by the material world is the prime
experience needed to initiate individual life into meaning.
In the first awakening, elements arouse us to an awareness
of *their* nature, but also to *our* very existence. Substances
summon us. With our attention captivated, we feel impelled
to act. We are motivated by the will to power, which, in
turn, repeating Bachelard's language from *The Right to
Dream*, "awaken[s] in us primitive acts, primary drives, the
imperious joy of ordering the world." By extracting images
from engagement with matter and then projecting them,
the imagination awakens, moves, and seeks to transform
the nascent, unawakened psyche. Bachelard defines the will
to action as the impetus for forming and expressing identity.

Important to our identity is the reciprocity between
our daytime reveries, nighttime dreams, and physical work,
for "matter is for the worker a condensation of dreams of
energy." Also in the first earth book, *Earth and Reveries of
Will*, Bachelard develops those dreams of definite action that
he designates as "reveries of will." The idea of dreaming, or
of reverie, as being linked to the will to action, may be one

of Bachelard's most original premises. Generally the acts of willing and dreaming are conceived as being opposite urges. Bachelard makes a good case that there would be no effective will without the impetus of the dream behind it.

With any endeavor in which we are fully engaged we become happily creative and feel like masters of our destiny. In the following comment concerning matter that is formed by craftsmanship, Bachelard explains how our choice of work and the energy we put into it reveals our intentions:

> If I were a psychiatrist with patients to attend, instead of a mere philosopher, hoping to learn from books alone, I would encourage the patients in my charge to free-associate, beginning with images representing the major human occupations. For it seems to me that in so doing one would discover not only the associations of ideas, but the *associations of energies*. One would have easy tests for a patient's courage or assertiveness. One might quantify and classify the very will to live, tabulate muscular desire, or reluctance to take action in the world. An individual's grasp of reality or lack thereof is more usefully determined through the use of such imagery than through any more conceptually oriented examination. Images, in effect, are less grounded in socially constructed meanings than are concepts and so more apt to reveal the solitary individual, the individual will. Tell me about work at the forge as you imagine it, and I'll tell you how much heart you put into your work.[30]

Bachelard, though an avid scholar and reader, constantly extols the value of physical work for the joy it brings in expressing oneself. We feel pleasure, and even awe, when any creative work is successfully accomplished, and Bachelard would argue that the secret to any task, even the most mundane housework, is to do it mindfully. He reiterates this sentiment in saying that there is a truism linking liberation of the soul to work: "All *creative acts* involve an overcoming of anxiety. To create is to unknot anxiety. At each new *challenge* we stop breathing. A condition of *working asthma* thus exists at the heart of every apprenticeship . . . But work is its own therapy, with benefits that carry deep into the life of the unconscious."[31] Although we might argue that some work is mindless, Bachelard insists that most if not all can be turned into creative endeavor—even the polishing of furniture.

In *The Right to Dream*, Bachelard insists that "everyone who labors dreams a cosmic dream." The world presents jobs waiting to be done. The destiny of work, though issuing out of the dream, is present in our bodies. Nothing that we have fashioned is a waste of time for, as Bachelard says in *Earth and Reveries of Will*, "The earthen objects we work return an echo of the inner forces we expend on them." Any activity in which we engage our bodies, especially our hands, calls forth imagination and participation: "A person's whole being comes to life when the hand takes control of matter."

This place is only one of many where Bachelard amplifies and expands the image of the human hand. In molding clay or in making dough, which he calls feminine labors, or in any involvement with matter where we use our hands, the will comes into play: "Hands which remain

undifferentiated when working dough—a feminine labor—
each take on their own particular dynamic value in work
of the third kind, against solid matter. This is why hard
substances serve as great educators of the human will,
regulating the level of nervous stimulation in our work,
and even contributing to our sense of virility."[32] Bachelard's
contention is that in attacking solid, hard matter we gain
confidence. In *The Right to Dream*, he alludes to the clenched
fist, which is linked to toil and needs toil, as "the digital will
. . . a will to build." Being endowed with a hand, we can
dream of holding the world in it.

In our attempt to master the earth and fashion it to our
liking, we learn patience, but also anger and how to invest
our energies. In *Earth and Reveries of Will*, Bachelard shows
us that anger can be a liberating force: "From unhappiness
at feeling stuck, feelings of anger rise in dialectical response,
and with them a sense of liberation." Bachelard encourages
a filament of anger as the best prod to productivity. He
comes to this startling conclusion about any endeavor: "As
by a provocation, the world is created through anger. Anger
lays the foundations for dynamic being. Anger is the act by
which being begins. However prudent an action may be and
however insidious it promises to be, it must first cross over
a small threshold of anger. Anger is the acid without which
no impression will be etched on our being. It creates an
active impression."[33] Isn't it reassuring that we are allowed a
little anger as not only acceptable but necessary? It is often
the required emotion that stirs us into engagement and
therefore becomes self-defining.

The imagination takes its cues from the world, but goes
beyond, creating worlds anew. In addition, imagination,

ever expansive, is like a giant sieve in which we sort out the kernels of our own authentic being. In *Earth and Reveries of Will*, Bachelard champions the process of participation in imagination's tendency to carry us beyond what is thought of as "reality," which, though seemingly filled with certitudes, he knew from his scientific studies, was only appearance: "It is in transcending reality that imagination reveals to us *our own* reality."

Bachelard extols imagination even in its exaggerated expansions, when it extends all the way to what he calls "the ir-real," or surreal. I think this affinity is what endeared him to the Surrealist painters, although he was not as devoted to the unconscious image as they were. He makes one of his witty inversions on the necessity of the unreal as well as real: "The way in which we escape reality gives a clear indication of our inner reality. A person deprived of the *function of the unreal* is just as neurotic as the one deprived of the *reality function*. It could even be said that difficulties with the function of the unreal have repercussions for the reality function. If the imagination's function of *openness* is insufficient, then the perception itself is blunted. We must find, then, a regular filiation between the real and the imaginary."[34] Perception in everyday life profits and even keeps us in balance by maintaining close lines of connection to the unreal and the dream world, as in *Water and Dreams*: "Real life is healthier if one gives it the holiday in unreality that is its due."

Bachelard celebrates the faculty of imagination in singular fashion. And it is not just that seeing the world through our imaginative glasses prettifies our vision. "The imagination is an 'accelerator' of the psyche," or, in other words, the imagination both amplifies and magnifies. Then, too, the imagination is what allows us to envision the future: "By the swiftness of its actions, the imagination separates us from the past as well as from reality; it faces the future. To the *function of reality*, wise in experience of the past, as it is defined by traditional psychology, should be added a *function of unreality*, which is equally positive . . . . Any weakness in the function of unreality will hamper the productive psyche. If we cannot imagine, we cannot foresee."[35]

Earth and Reveries of Will directs specific attention to another vital subject—the will—and allows us to examine its use in a most revealing way. In studying psychology one becomes aware that, with the exception of Alfred Adler, the whole discipline has little to say about will, which Bachelard considers a primal human trait. Perhaps this situation is a result of the way will traditionally has been treated—as a spirit quality—and therefore more in the realm of philosophy than psychology, which by its name is primarily concerned with the word of the soul. In psychology, will often seems to equate with ego. Bachelard has a much more expansive definition, rescuing will's reputation and providing a broader appreciation than is habitually attributed to it. His classification in *The Right to Dream* embraces the alignment of meaning and order in one's life, for will is "an invitation to action, that calls for the intervention of man in the world, that channels the chaotic forces of creation." Faced with an incomprehensible confusion of material forces at large in the world, our will commences its job of defining us. These choices serve the impersonal urge of furthering our physiological growth and pushing us along the road of individuation.

The effort we expend in challenging the material world is productive because, in actuality, we learn more through opposition than through affinities, which do not galvanize us into response. In *Earth and Reveries of Will*, Bachelard insists: "The *human* being is revealed as the being *opposed* to things, not siding *with* things but standing up *to* them." The struggle to resolve ambivalence in our lives is endless. Bachelard teaches us not only that this is fruitless, but indeed that even to seek unities is to stultify imagination. He assures us it is perfectly plausible to live with ambivalence and uncertainty. However, we must expect that the will, awakened to a sense of power, always carries a kind of naiveté. We imagine we are capable of unlimited possibilities, because imagination is expansive and naturally exaggerates. By a grounding in images of power, we discover a royal road to power in *The Right to Dream*: "The will to power needs images; the will to power is thus matched by an imagination of power."

Imagination and will have always seemed in opposition to one another, but not since Bachelard's breakthrough concept joined them in the dream's desire to work with matter through engagement with image. To link will and being-in-the-world provides a healing combination of living matter with soul and spirit, so often disconnected.

Bachelard surprisingly also relates will to contemplation. In *Water and Dreams* he says, "Contemplation is not opposed to will but rather follows another of its branches, participating in the will for beauty which is an element of will in general." Because we have a natural desire to understand our surroundings, we begin by opening our eyes to the world. This type of curiosity draws us into seeing more acutely. Moments of contemplation, prior to the demands of making and doing, are rewarding. His claim is that "Aesthetic contemplation alleviates human sorrow for an instant by detaching man from the drama of will. This separation of contemplation from will eliminates a feature that I would like to stress: the will to contemplate. For contemplation also gives rise to a kind of will. Man wants to see. Seeing is a direct need."[36]

Bachelard takes psychoanalysis to task, something he often does, for failing to give imagination its essential and unique place in the psyche. In *The Poetics of Reverie* he says: "It might perhaps be simpler if we were to follow the tried and true methods of the psychologist who describes what he observes, measures levels, and classifies types—who sees imagination being born in children without ever really examining how it dies in ordinary men."[37] Bachelard argues in *Air and Dreams* that the problem of psychoanalysis is that it "neglects the problem of the imagination, as though the imagination were unproductive time-off from a persistent, affective occupation." It is more than mere classification of observations. At the time of Bachelard's first venture into the elemental imagination, psychology, in following Freud's newly described interpretation of dreams, tended to reduce each symbolic act in a dream to a single underlying motivation. Bachelard accuses psychoanalysis of often handling "a knowledge of *symbols* as though they were concepts." For example, in dreams, when one dreams of flying or falling, psychoanalysis "assigns a definite meaning once and for all to a particular symbol." Bachelard points out in *Water and Dreams* that "*oneiric pleasure* is satisfied by making the dreamer *fly*," but this acknowledgement doesn't go far enough to explain the aesthetic and dynamic appeal of the act of flying. To explain the gracefulness and the dynamic lift of images of flight requires further exploration. Every "gracefully curved trajectory must be followed by a responsive inner movement." Bachelard offers a new perspective on the aesthetics of grace. Here he ascribes how even the emotion of love and the experience of ascending are closely related:

> Every graceful line, then, discloses a kind of *linear hypnotism*: it guides our reverie by giving it the continuity of a line. But beyond the imitative intuition that obeys, there is always an impulse that commands. To one who contemplates a graceful line, dynamic imagination suggests a most unlikely substitution; you, dreamer, you are grace in motion. Feel *graceful power* within yourself. Become conscious of being a reservoir of grace, of being a potential for breaking into flight. Understand that within your will are scrolls coiled like the new leaf of a young fern . . . . The dream of flight is the dream of a *seductive* seducer. Love and its images cluster around this theme. By studying it we will see how love *produces* images.[38]

This quotation demonstrates how Bachelard takes one inside an image, describing how it feels to participate in an image. He is never a detached observer, experiencing through his head, or merely his eyes. He carries us deep into the feeling of the movement itself. The joy of ascension is not only akin to the pleasure of a soul filled with love but also is satisfying to the nighttime dreamer. In *Air and Dreams* Bachelard urges: "Those who study sleep . . . should pay attention to the nocturnal experience of oneiric flight." To be lifted up counteracts the innate fear of falling. This observation explains why "For certain souls whose nocturnal activity is very powerful, to love is to fly; oneiric levitation is a more profound, more essential, less complicated psychic reality than love itself. The need to become lighter, to be freed, to take great freedom from the night, appears as a psychic destiny and as the very function of normal nocturnal activity: of a restful night."[39]

One of his most meaningful statements contradicts the usual notion of image-making as a gathering together of perceived images. When trying to explain how Bachelard regards the imagination as concrete and really real but also always mobile, expanding, inclusive, and never closed or completed, I like this statement from *Air and Dreams*:

> We always think of the imagination as the faculty that *forms* images. On the contrary, it *deforms* what we perceive; it is, above all, the faculty that frees us from immediate images and *changes* them. If there is no change, or unexpected fusion of images, there is no imagination; there is no *imaginative* act . . . [40]

Air

Without realizing it we have slipped over into the area of the third element—air. Bachelard calls this element "masculine" and attests that one of its chief characteristics is the love of freedom, not only as a philosophical ideal but as a breaking away from restraints. We unconsciously make an affiliation with one of the elements. This choice shapes our psychology and, in building a foundation, shapes the way we respond to future experiences.

It's not only that we experience the pleasure of a gentle spring breeze, a beautiful vineyard laid out in neat rows, looking at the setting sun over the ocean, or sitting by a fire, but it's that these engagements with matter also arouse our desire to participate. Bachelard maintains that the contrasting feelings of opposition also draw us into efficacious engagement. He suggests in *Water and Dreams* that the pride we feel when able to counter an adverse material element gives dynamic unity to a being, "a proud victory over an adverse element." The conqueror, feeling victorious, experiences an *élan vital*, a renewal of energy. All resistance awakens the will to conquer and the same wish: "In the realm of will, there is no distinction to be made between things and men." Man is an integrated being. He has the same will against all adversaries. The world is our provocation, shaping our entire value system.

Bachelard makes a strong case for the philosopher/ poet Nietzsche as a lover of high, cold air. He explains why the other three elements are not as active in his imagination as that of air: "Nietzsche is not a poet of *earth*. Humus, loam, cleared and ploughed fields do not provide him with images," and, too: "Nietzsche is not a poet 'of matter.' He is a poet of action." Nor is Nietzsche a *water* poet. "There are, of course, water images. No poet can do without liquid metaphors. But, with Nietzsche, these metaphors are ephemeral. They do not determine *material reveries*."[41]

THE SEDUCTION OF MATTER

Bachelard instructs us to pay attention to the parts of speech that each poet chooses: "We must determine the exact weight of every adjective if we wish to understand the metaphoric life of language."[42] With his joy in freedom, the element of air proves to be the most dynamic for Nietzsche: "For Nietzsche, in fact, air is the very substance of our freedom, the substance of superhuman joy. Air is a kind of matter that has been mastered, just as Nietzschean joy is human joy that has been mastered. *Terrestrial* joy is richness and weight—*aquatic* joy is softness and repose—*igneous* joy is love and desire—*aerial* joy is freedom."[43] All poets rely on a multitude of elemental images, but in his many books Bachelard makes examples of a number of different poets, each being especially moved by a singular, dominant image. I have tested out and written about his theory in William Butler Yeats's poetry and found that fire imagery is by far the most frequently used element in Yeats's canon.

With regard to the the imagination of air, we can understand that it is, as Bachelard says, "very thin matter." That is why density is not an issue, or is of less importance, than it was with elemental earth. Here it is the degree of the ability to move about freely and in what direction that matters. His attention is focused on such subjects as flight, wings, the imaginary fall, the sky, clouds, the wind, and silent speech. He explains that every element has its own characteristic dynamism, what gives it its energy. With air, it is movement: "Every element that the material imagination enthusiastically adopts prepares a special sublimation, a characteristic transcendence, for the dynamic imagination . . . . We will see that aerial sublimation is the most typically

49

discursive sublimation, the one whose stages are the clearest and the most regular . . . . The adjective most closely associated with the noun *air* is *free* . . . . On the other hand, air offers a distinct advantage when one comes to the dynamic imagination. With air, movement takes precedence over matter. In this case, where there is no movement there is no matter. The aerial psyche will allow us to develop the stages of sublimation."[44] In other words, with air, we can follow more easily the vector as imagination unfolds and expands. Always imagination works by association. We might note here that Bachelard's use of the word "sublimation" differs from the traditional psychological one, which holds that sublimation is a covering over of unwanted or unacceptable material that we have buried.

The following statement about aerial images demonstrates Bachelard's way of defining both images and, in this case, the element of air. He asks the reader's permission to once again consider "the one characteristic that I want to examine in aerial images: their mobility, comparing this external mobility to the mobilism that aerial images induce in us. To put it another way: images, to my way of thinking, are psychic realities. Both at the time of its birth and when it is in full flight, the image within us is the subject of the verb, to imagine . . . . In human reverie, the world imagines itself."[45] He turns inside out our normal ideas of consciousness. It is not that we see the world and have perceptions first and then imagination plays on these perceptions. He emphasizes that images are primary psychic realities. In experience itself, everything begins with images. This means that images are *not* primarily visual, as we automatically assume, nor are we simply passive observers of them. Aerial images induce a sense of escape and freedom when we mutually participate in them.

Bachelard's attention to the vertical axis can help with how we give value, positively or negatively, to any endeavor. We grow *up* but we fall *down*. Bachelard explains that valorization is always in terms of up and down, or in such words as depth and height. For example, in the circles of the *Inferno* of Dante's *Divine Comedy*, in the passage downward, air becomes more and more scarce for the pilgrims Virgil and Dante. With the increasing depth, the sins represented become more malicious, and there is less freedom to move. This image echoes Bachelard's insistence that air equates with freedom.

Bachelard affectionately lays to rest any notion that valorizing the world, poeticizing it, is a passively submissive act. He assures us that the poetic act is still essentially the will to creativity, and creativity is power.

On this subject, I would like to interject an observation from a chapter in my book, *The Bonding of Will and Desire*. I wrote the following about the attraction to the material elements that often reveals personal inclinations:

> Each encounter with matter, each material alliance rouses the desire to touch, or to conquer the element, and therefore evokes responsive imaging. Distinctive complexes develop in association with the imagination of matter in each traditional group. By matching these qualities with individual personality traits,

we obtain an amplified portrait of elemental proclivities, a mirror in which to view our reactions. To oversimplify a bit, we could say that those who are drawn to water respond to the always changeable, always profound depths of experience. They are reflective. Those drawn to air value a sense of freedom and lightness, leading to upward, aspiring thinking. Those who associate with the earth veer toward stability, endurance, practicality, and determination. Those who respond to fire embody liveliness, emotional intensity, and the heroic. All of us have moments of strong affiliation with each element, but we have a dominant inclination that remains constant in our interaction with matter.[46]

Fire

Our final element to consider—fire—is the most intense, and the one with the most extreme extensions. As always, the attachment to an image, in this case one of fire, indicates inner proclivities. Bachelard makes this connection in *Psycholanalysis of Fire*: "Fire is more likely to smolder within the soul than beneath ashes." Fire carries the most extreme opposites—from the warmth of the hearth to the hellfire of damnation.

The Psychoanalysis of Fire, the first book on an element and the beginning of what was to develop into a series on the imagination, had as its purpose the dispelling of the illusions concerning fire. Bachelard maintained that scientific thought has never adequately dealt with the phenomenon of fire. In the middle of the book, he makes a major and permanent shift in his interest. Henceforth he will pursue studies in the way imagination colors all perception. In addition to his books explaining the connections between scientific thought and the living human being, he will develop new theories of the rich, imaginal life. After this initial book, Bachelard wrote two other books on elemental fire: *The Flame of a Candle*, on a single image, and the book that he was working on when he died, *Fragments of a Poetics of Fire*, in which we find this: "A complete psychology will require of us that we live at both poles of our androgynous being, able thus to experience the fire both as violent and reassurance, sometimes as the image of love, sometimes as the image of anger."[47]

Elemental fire, in the imagination, equals strength, ardor, and courage. For example, when Achilles is on fire with desire to avenge Patroculus's death in the *Iliad*, his body is described as being surrounded by fire. In its slower aspects, fire is internal digestion with the gradual fire of transformation. In a note Bachelard explains: "Our internal organs are so many hearths. A whole range of idioms which make use of fever as a metaphor are used in describing human instincts. An existentialism of the senses—could any other kind exist?—has need of an infra-language. Fire must be felt to be a valuable possession safely smoldering beneath the ashes. The attraction in buried fire speaks through a thousand torrid dreams. Treasures burn, as do we, with covetous desire for them."[48] We describe ourselves as being on fire for some material goods in such statements as "being on fire with ambition." Or, "on fire with desire" for another, as in Yeats's poem "The Mask," in which the protagonist is attempting to reach beyond the attraction to the persona, or mask, of the lover, but concludes: "What matter, so there is but fire / In you, in me?"

Bachelard continues:

> Once we are willing to experience fully the prodigious varies of images of fire, flames and glowing embers, we become aware that we ourselves are living fire. The most important lesson to be learned from this psychology of the experience of fire [le feu vecu] is the importance of opening ourselves to a psychology of pure intensity, the intensity of being. Once it has been seen that fire's essence is the essence of intensity, it should be possible to claim the reverse as well. Our being rises and falls, brightens and darkens, never at rest and never stable, living always in a state of tension. Fire is never still. It lives on even as it sleeps. Our experience of fire is always a matter of heightened existence. Yet, thanks to these images of imagined intensity, one is spared the too-sudden brutality of intensity experienced in the flesh.[49]

Next, Bachelard addresses fire as a purifier and deodorizer, ridding matter of its heavier and grosser qualities. Finally, he completes the cycle by relating fire to the light that enlightens the mind. As usual, we have some examples of culture complexes. He begins with the Prometheus complex, taking its name from the pride and defiance in the Titan's act of stealing fire from Olympus and giving it to the human race. In *Fragments of a Poetics of Fire* Bachelard left many unfinished notes. Here is one of the most interesting ones: "The bearer of fire is a bearer of light, the light of spirit, of metaphoric clarity, of consciousness. What Prometheus stole from the gods to bestow upon humanity was consciousness. The gift of fire-light-consciousness opened the door to a new human destiny. This destiny of consciousness, of spiritual knowledge, however, is not an easy one."[50]

The Prometheus complex is one that holds much interest for our particular time, as we have become Promethean in so many ways, denying any attribution to a higher realm of being or higher authority. Many no longer consider themselves as God's children. This complex allows us to think of ourselves as entirely independent. You will recall that Prometheus, a Titan and a member of the generation preceding the Olympians, stole fire from the gods and gave it to mankind, a great gift for humanity. Zeus punished him for this brazen act by confining Prometheus to a rock, where every night an eagle bore down on him and ravaged his liver. Bachelard carries the mythology forward into current life by explaining the manifestations of not thinking we owe anyone anything. The will-to-intellectuality, one natural aspect of an upwardly mobile imagination, easily turns into a Prometheus complex. Describing the complex as "the intellectual mastery of fire," Bachelard recognizes that "there is in man a veritable will to intellectuality." Thus this complex involves "those tendencies which impel us *to know* as much as our fathers, more than our fathers, as much as our teachers, more than our teachers."[51] It is the desire not only to surpass one's father but also, in essence, to defy the gods, to steal their fire and their creative energy.

In the poem "Prometheus Unbound," Percy Bysshe Shelley accurately foresaw Prometheus's acceptance as the secular god in Western intellectual recesses. Indeed, everyone in our culture today shares some degree of Promethean attributes. The Prometheus complex reigns in the secular environment of "exhausted spirituality" (Solzhenitsyn's phrase), ever generating the lamentable belief that the gods are passé. In Bachelard's view in *Fragments of a Poetics of Fire*, "All educated persons organize their thinking around one figure of Prometheus or another . . . . There is a personal Prometheus for everyone." This willful deity is the prototypical American hero, always probing for new realms to conquer, quickly dismissive of past burdens or debts of gratitude. Prometheus's special gift is peering into the future, and he finds it teeming with opportunity. He *is* the future tense governing our projections. When we say, "I *will* learn," we are projecting our Promethean urge. Positively, Prometheus is the will to new knowledge, the spur to experimentation. He is also our power overdrive, our will to exceed limits, our denial of all restraints, our defiance of the gods in the cause of self-aggrandizement, our glorification of what it means to be human. Bachelard makes this important connection: "The bearer of fire is a bearer of light, the light of spirit, of metaphoric clarity, of consciousness. What Prometheus stole from the gods to bestow upon humanity was consciousness."[52]

Bachelard focuses attention on how the Prometheus complex tempts us to ignore restrictions: "The many figures of Prometheus inherited from past myths and cultures have rooted themselves in us, making psychological techniques of *self-transcendence* possible . . . . But if these imposing figures of Prometheus are to have any psychological effect on us, they must be experienced as attempts—or, better yet, temptations—to transcend our own natures, to experience the human, more than human."[53] In his last notes for *Fragments of a Poetics of Fire*, Bachelard is particularly concerned to further develop the figure of Prometheus, writing that "The hero Prometheus is a symbol of *constructive disobedience*,"[54] and adding that "Promethean imagery in poetry always has the psychological effect of *uplifting* human nature."[55] Among scholars, the late Donald Cowan emphasizes the positive side, that "The Promethean fire is imagination, not simply skills, or reason; and imagination is the power from a divine source whereby matter is permeated with spirit."[56] If we count our blessings as gifts, we avoid the hubris of self-inflation.

Another vitally interesting complex, for obvious reasons connected to fire, is the Empedocles complex. It takes its name from the philosopher who threw himself into the annihilating fire of Mount Etna in Sicily to prove that the soul never dies. Bachelard says in *The Psychoanalysis of Fire*: "Empedocles chooses a death which fuses him into the pure element of the Volcano." We may think of his act as courageous or foolish, but Bachelard calls him a hero because he is willing "to bear the weight of his hero's destiny alone. Empedocles in this respect is the true hero of being face to face with essential nothingness. We have reached the heights of heroism here, heroism both gratuitous and causeless."[57] Bachelard adds a note that Empedocles is said to have left only a sandal behind him in the world. Why does the abyss seem ever threatening, but also compelling? Bachelard answers: "Great and small have each their own abyss, as does each element. Empedocles, for example, was lured by the abyss of fire."[58] In the Hoffman complex, he explores the liquid fire of the punchbowl, or of any alcoholic substance.

In *The Poetics of Fire*, Bachelard includes extensive additions and notes on the Empedocles and Prometheus complexes, thoughts that he began exploring twenty-five years earlier in *The Psychoanalysis of Fire,* but that now are more closely aligned to Jungian psychology. Much of the new material concerns the image of the phoenix—the legendary bird that arises anew from the ashes. Bachelard's note says: "The phoenix is quite properly, the symbol of eternity." And indeed, this image provided an alternative title to this final manuscript—*The Poetics of the Phoenix.*

Bachelard expands our pleasure by calling our attention to the simple joys of awareness—to the colors, sounds, and textures of our surroundings. By our response, we confer attention on all animate and even inanimate objects and, in the same act, magnify our own sense of well-being. To sum up our legacy from Bachelard, as he states in *Earth and Reveries of Will*: "Matter is a revelation of being." Let us adopt the Bachelardian stance: Be open to the things of this shining world.

CHAPTER FOUR

ALCHEMY, THE SENSES, AND IMAGINATION

Emotion is the moment when steel meets flint and a spark is struck forth, for emotion is the chief source of consciousness.

In this chapter, we will see how Bachelard's engagement with the senses enables alchemical reflections between inner and outer, microcosm and macrocosm, the cosmic and the practical. Three of the senses that we are concerned with here—smell, taste, and sight—have their expression through the face. And warmth—not as a temperature reading but as a subtle quality of the heart—is also reflected in facial expression, how we face the world. I am going to use the framework of alchemy, specifically the four classical elements of earth, air, fire, and water in relation to the senses, to deepen our awareness of various ways of relating to the world. Physiological data that comes from a chart in a doctor's office is of use mostly as a springboard for understanding the psychological and spiritual aspects of being; it is our living experience that we are primarily concerned with here.

The alchemical way of viewing the world is one that

recognizes likenesses between life on the macrocosmic level and on the microcosmic level. This approach translates into seeing ourselves as both individuals and as part of the cosmic order, seeing ourselves in the world and the world in us. Bachelard gives us many clues as to how to accomplish this feat. Since the early 17th century, at the beginning of the scientific age, the intellectual direction in most Western nations had been pointing toward the further and further abstraction of humankind from the natural world. In the post-modern period, which roughly began in the 1960s according to some authorities, a counter-movement began. A few thinkers began seeking ways of entering into a more whole, all-inclusive view. Bachelard died in 1962, but he was already ahead of his time in countering straight, linear thought and strict rational causality. What we will be especially concerned with here is the modern habit of the splintering and division of human faculties. Bachelard teaches us to appreciate our inner world and the outer environment and the interplay between both. He teaches us to hold to the small act of noticing. Henry David Thoreau expresses much the same: "It's not what you look at that matters, it's what you see."

Our typical separation of the senses—and of the elements—provides tools for broadening our sensitivities to the material world, but none of the senses or elements are absolutely isolated. Tasting, for example, also is connected to smelling. When we have a head stuffed with a cold and can't smell or taste anything as acutely, we can easily see how the senses impinge on each other. For the sake of argument we are temporarily isolating them here to better understand them.

ALCHEMY, THE SENSES, AND IMAGINATION

Air and Smell

As we have seen, Bachelard classifies the four elements known to antiquity—earth, air, fire, and water—in his own unique way, designating them "hormones of the imagination." In other words, these elements excite our imagination in profound ways, stimulating the gathering of collected images, and thereby accumulating multiple associations. Bachelard explains that this classification helps in the inward understanding of the reality that is dispersed among multiple forms: "They [the elements] activate groups of images. They help in assimilating inwardly the reality that is dispersed among forms. They bring about the great syntheses that are capable of giving somewhat regular characteristics to the imaginary. Imaginary air, specifically, is the hormone that allows us to *grow* psychically."[1] Bachelard suggests that air is the element that causes us to stand tall, to reach for heights or for the sky, to be inspired.

Phenomenologically speaking, air is clear and pure, but mysteriously carries smells of a complex nature. Animals, of course, have a much more highly developed sense of smell than humans. They can distinguish other animals from quite a distance. Our human sense of smell grows less intense with age—there are even indications that the loss of the sense of smell is one of the indications of the onset of Alzheimer's disease. Air wafts the smells, both pleasant and disagreeable, that we immediately recognize. What brings more pleasure than the smell of newly-cut grass in the rain, or cookies baking in the oven! Much of what we taste also depends upon smell. And smells cause us to relive memories—recall the vivid description of baked madelaines in Proust's *Remembrance of Things Past*. Unpleasant smells also remain lively in the imagination. Shakespeare's couplet in his Sonnet 94 about the smell of decay in flowers that were once sweet-smelling, or, by analogy, someone who was once virtuous and turns evil, always strikes home with me: "Lilies that fester smell far worse than weeds."

Water and Taste

Alchemically, water makes everything less hard, less brittle, less stringent. It is the easiest element to contemplate with our bodies, because the body is 94% water. Drinking pure, clean water is a very enjoyable experience, and one that we need to take time to relish. Wouldn't it be a terrible suffering not to have access to clear, sweet water? We take it for granted, but in many parts of the world today, the lack of it has become a main concern. The element of water is a good place to begin with our association of *within* and *without*.

We will want to consider not only pure water, but refine this awareness by thinking of water in solution with salt, giving us salt water. Outwardly, the salt water of the oceans covers the earth. But coming from within, our tears and our sweat are very salty, as we know from tasting them as they run down our face. Salt preserves things inwardly, in memory. Sometimes we would prefer to forget our blood, sweat, and tears, but perhaps they are preserved in our salt-watery imagination of the past. It may be that we are meant to recall these defining moments. Don't they define us as unique individuals with our own special experiences?

Sometimes we laugh so much that we cry water as salty tears. Both the laughter and the tears relieve tension.

Salt is a product of the drying out of the seas of the earth. We say, "He or she is the salt of the earth," which means that he or she is truthful and courageous. Salt is one of the important ingredients of the alchemical reactions, which we will come to presently. According to Theophrastus Paracelsus, the most famous alchemist of the 16th century, saltiness is one of the four kinds of taste—sour, sweet, bitter, and salty.

Water carries many associations. As I mentioned before, its most constant association is with purity or purifying. "The sweetness of water impregnates the soul itself," according to Hermes Trimegistis. In *The Hermetica* can be found these lines: "An excess of water makes the soul sweet, affable, easy, sociable, and pliable." Hermes sounds so like Bachelard. In classical times, some considered that fish were composed of water, while mankind was composed of water and earth. Others insisted that man was composed of all four elements. *The Hermetica* refers, interestingly, to the doctrine of Posidonius concerning life after death: "Posidonius seems to have said that men living on earth are composed of all four elements, and when a man dies, he becomes first a 'hero,' composed of water, air, and fire, and dwelling in the lower atmosphere; then a 'demon,' composed of air and fire, and dwelling in the upper atmosphere; and (in some few cases) finally a god, composed of fire alone, and dwelling in heaven."[2]

Just as we have been doing here in relating water to tasting, we have seen in Chapter Three that Bachelard relates water (an element) to seeing (a physical attribute).

By seeing and revealing ourselves, water causes us to reflect. If, though, we dwell excessively only on our reflection, to the extent of ignoring others, we become like Narcissus. Here we will appreciate how, with his method of circular approach, Bachelard weaves all the complications of water together in *Water and Dreams*. It is not only the pure, necessary liquid that sustains life, or that washes away sins or impurities, it is also nature's mirror: "The world wants to see itself . . . the true eye of the earth is water."

The element of water teaches us to appreciate and to understand sound and rhythm, Bachelard insists. Here again as he so often does, Bachelard makes an acute observation: "Human language has a *liquid quality,* a flow in its overall effect, water in its consonants. I shall show that this *liquidity* causes a special psychic excitement that, in itself, evokes images of water . . . Thus water will appear to us as a complete being with body, soul, and voice. Perhaps more than any other element, water is a complete poetic reality."[3]

Of water images, Bachelard claims: "Images whose basis or matter is water do not have the same durability and solidity as those yielded by earth, by crystals, metals, and precious stones. They do not have the vigorous life of fire images."[4] Elemental fire and air are masculine, claims Bachelard, while water and earth are feminine. Perhaps this is why we call her Mother Earth. In the alchemical process, liquefaction opposes combustion, or water opposes fire. Again Bachelard makes this distinction: "We shall also see how profoundly maternal the waters are. Water swells seeds and causes springs to gush forth. Water is a substance that we see everywhere springing up and increasing. The spring is an irresistible birth, a continuous birth. The unconscious that loves such great images is forever marked by them."[5]

Earth and Seeing

We have been dwelling on the relationship of water—first to taste, then to to seeing, being seen, and speaking—and have naturally slipped into elemental earth and the faculty of seeing. Earth is not only "the right place for love," to borrow from Robert Frost in "Birches," it is also the right place for seeing. Bachelard designates the earth as the place where we learn about vastness: "If images are accorded their rightful place in the workings of the mind—prior to thought—then one cannot fail to recognize that the fundamental image of immensity is *terrestrial*. The earth is immense, vaster by far than the sky, which merely functions as its dome or roof."[6]

Earth's horizon beckons us to stand up and move forward, to see and explore. In *Earth and Reveries of Will*, Bachelard develops this idea of wishing to see the world more fully:

> This notion of *imagined immensity* helps us to understand how an infinite variety of spectacles can be integrated in a unified vision of the earth. Nomads, despite their wanderings, are always at the *center* of the desert, at the center of the steppe. No matter which way one turns one's eyes, diverse objects capture one's attention, and yet a force of integration knits one's field of vision into a shared circular panorama with the dreamer at its center. Such "circular" vision

embraces the horizon fully. There is nothing abstract about this sweeping view over the broad immensity of the plains. The psychological reality of the panoramic gaze is available for each of us to experience with intensity if only we take pains to look. In sweeping the gaze round the horizon, the dreamer takes possession of the whole earth.[7]

By opening our eyes and simply giving ourselves to gazing, we are exhilarated by the earth's beauty:

The dreamer *dominates* the universe, and to neglect this peculiar and banal form of *domination* would be to leave out an important element in the psychology of contemplation. From its very first contact with immensity, the contemplative gaze would seem to come into a sudden mastery of the universe. Apart from any philosophical reasoning, the sheer restful tranquility of the open plain joins eye and world together in contemplation—a seeing being, who enjoys seeing, who finds *beauty in seeing*, before whom there stretches the immense spectacle of the world's immensity, a world *beautiful to see*, whether composed of infinite sands or cultivated fields. And those who contemplate in this fashion can share credit for the beauty of their vision.[8]

Bachelard forever emphasizes themes of the four basic elements of alchemy (fire, water, air, and earth). He has this observation: "Let us take the least sensual of the sensations, the visual, and see how it becomes sensual." We may confuse the two words, "sensory" with "sensual." Sensations are highly mental data. Bachelard makes this distinction between the proper classifications of sensual values in relation to sensory ones. "I believe that the doctrine of the imagination will be clarified only by a proper classification of sensual values in relation to sensory ones. Only sensual values offer 'direct communication.' Sensory values give only translations. Confusing the 'sensory' and the 'sensual,' writers have claimed a correspondence among sensations (highly mental data) and have therefore failed to undertake any study that considers poetic emotion in its dynamics."[9] It is when sensory data becomes a sensual value that it has its most profound effect upon us.

Bachelard is thus inviting us to reawaken aspects of our experience that pre-date the heavily *sensory* nature of enculturation. Feral animals generally have a more sensitive sensory physiology than the human race. They are super-sensitive to sound, temperature, touch, vibration, electrostatic and chemical activity, and magnetic fields. Before a tsunami wave hits a beach, certain animals will run to higher ground. Others can sense seismic activity and subtle sounds—infra-sound frequencies in the range of one to three hertz, compared with humans 100-200 hertz range. Fish detect very low-frequency vibrations. Dogs and cats are sensitive to electromagnetic changes that precede earthquakes. It is possible that we too, following Bachelard's clues, can become receptive to a much wider spectrum of experience via pre-reflective *sensuality*.

An Elemental Reverie

Fire and Warmth

Warmth is not generally considered one of the senses, but it is certainly a quality of which we are very aware. Warmth is associated with being alive, with life itself. A few degrees of fever can make us feel quite a different person. When I take medicine, sometimes my temperature drops and I feel dragged down. I have always marveled that the body can maintain its constant temperature, even when the outside temperature drops below zero or climbs above 100 degrees. What a marvel the body is! I have always been fascinated that two different people, mates for instance, may prefer the thermostat at different temperatures. Just a few degrees can make a major difference in comfort. Maybe it has something to do with metabolism, but one of the small irritants of daily living can turn on how hot or cold a room should be, especially at night while sleeping.

Elevated heat in the body is associated with anxiety, anger, or agitation of some sort. We speak of "getting hot under the collar," or that something "burns us up." We get red in the face, or sweat when under emotional pressure. My mother used to warn me, saying: "What burns you up also burns you out."

In addition to these physiological attributes, I want to discuss the psychological feeling of warmth. Warmth of feeling toward another is connected to the heart. On Valentine's Day, the setting of one's heart on fire is frequently the motif. "Come on baby, light my fire" is the frequently repeated chorus of a well-known song about the feeling of excitement of sexual attraction. First, let us make a bit of separation between feelings and emotions. Though we feel emotional, feelings and emotions are not the same. Let me attempt to define them separately. Feelings put us in dialogue with the world. Emotions, while important, are more of a private reaction to our individual experiences. Michael Lipson observes the distinction between feeling and emotion: "As we begin to turn the rich, juicy quality of our emotional life away from our own sensitive souls, we can discover innumerable new feeling-experiences as that world impresses itself upon us. Every moment of experience could come to us with a new feeling-tone, if we met the moment with our full attention."[10]

Many believe that conscious awareness originates in the brain alone. However, recent scientific research suggests that consciousness actually emerges from the brain and body acting together.[11] A growing body of evidence suggests that the heart plays a particularly significant role in this process. To the extent that this is the case, warmth—and fire—play a fundamental role in consciousness.

Far more than a simple pump, as was once believed, the heart is now recognized by scientists as a highly complex system with its own functional "brain." Research in the new discipline of neurocardiology shows that the heart is a sensory organ and a sophisticated center for receiving and processing information. The nervous system within the heart (or "heart-brain") enables it to learn, remember, and make functional decisions independent of the brain's cerebral cortex. Moreover, numerous experiments have demonstrated that the signals the heart continuously sends to the brain influence the function of higher brain centers involved in perception, cognition, and emotional processing.

In addition to the extensive neural communication network linking the heart with the brain and body, the heart also communicates information to the brain and throughout the body via electromagnetic field interactions. The heart generates the body's most powerful and most extensive rhythmic electromagnetic field. Compared to the electromagnetic field produced by the brain, the electrical component of the heart's field is about 60 times greater in amplitude, and permeates every cell in the body. The magnetic component is approximately 5000 times stronger than the brain's magnetic field and can be detected several feet away from the body with sensitive magnetometers.[12]

The heart generates a continuous series of electromagnetic pulses in which the time interval between each beat varies in a dynamic and complex manner. The heart's ever-present rhythmic field has a powerful influence on processes throughout the body. An essay by Rolin McCraty, "The Resonant Heart," demonstrates that brain rhythms naturally synchronize to the heart's rhythmic activity, and also that during sustained feelings of love or appreciation, the blood pressure and respiratory rhythms, among other oscillatory systems, entrain to the heart's rhythm.[13]

The message I take from this research is this: whatever the emotion is, the key is not to "take it to heart." The heart should be a sanctuary—not a receptacle for the tumult of the marketplace. Nonetheless, from an alchemical perspective, strong emotions—including anger—have a part to play. Jung, for instance, values the fire of anger, giving this stimulus the credit for arousing consciousness:

70

Conflict engenders fire, the fire of effects and emotions, and like every other fire it has two aspects, that of combustion and that of creating light. On the one hand, emotion is the alchemical fire whose warmth brings everything into existence and whose heat burns all superfluities to ashes. But on the other hand, emotion is the moment when steel meets flint and a spark is struck forth, for emotion is the chief source of consciousness. There is no change from darkness to light or from inertia to movement without emotion."[14]

Bachelard echoes and extends Jung's perspective:

Hard matter is conquered by the angry hardness of the worker. Here, anger is accelerative. Moreover, in the realm of work, any acceleration demands a certain anger. But the anger involved in labor destroys nothing. It remains intelligent, illuminating the line of Vedic verse cited by Louis Renou: under the effects of *soma*, "Craft and anger come alive, *o liqueur!*" . . . Naturally such anger speaks. It provokes matter. It insults. It triumphs and it laughs. It ironizes. And it writes literature . . .[15]

Quoting Jacob Boehme, Bachelard adds, "We each experience the source of anger and bitterness at the origin of our lives, for if not we would not be alive."[16]

Fire, heat, passion, elemental emotion—all are dangerous. Yet in some fundamental, alchemical sense, all are needed for change and transformation. The word

71

element was first used by Empedocles, who believed that all matter, including man, was composed of some combination of the four elements. Titus Burckhardt defines the less obvious qualities of the elements: "If the elements are listed in order of their 'fineness' or 'subtlety,' earth comes lowest, and air highest. If, however, they are ordered according to the directions of their movement, fire occupies the highest place. Earth is characterized by heaviness; it possesses a downward tendency. Water is likewise 'heavy,' but also has the capacity of 'extension.' Air both rises and extends, whereas fire rises exclusively."[17] Paracelsus says of the elements: "Man consists of the four elements, not only—as some hold—because he has four tempers, but also because he partakes of the nature, essence and properties of these elements."[18]

The alchemical process, on either the metallurgic/chemical plane, or the soul/psychological plane, is all about transformation. The goal of the chemical approach involves the effort to transform base metal into the philosopher's stone, or alchemical gold, by extracting a pure elixir from a chaotic mixture. On the soul plane, the goal is to transform uninitiated man into a spiritual being. Symbolically, the alchemical process is divided into six steps. The first three—"ascending"—are "the lesser work." Burckhardt provides this illumination of the stages of spiritual transformation: "The first three stages correspond to the 'spiritualization of the body,' the last three to the 'embodying of the spirit,' or the 'fixation of the volatile.' Whereas the 'lesser work' has as its goal the regaining of the original purity and receptivity of the soul, the goal of the 'greater work' is the illumination of the soul by the revelation of the Spirit within."[19] In an elementary sense, then, we could say that the first three stages involve ascending toward a more heavenly realm, through distilling or burning away of what is unnecessary. Then, after gaining the knowledge of Spirit, the subsequent steps involve "descending"—returning and investing the earth with embodied Spirit.

Bachelard rarely speaks directly about alchemy, though when he does it is with admiration for the imaginative processes that are engaged. His interest is not in metaphysics per se, or the end result of the alchemical process as a refinement of the human soul, or its extraction from dross matter. He loves matter. In his own unique way, Bachelard is enacting a kind of enhancing, an enlivening, even a spiritualization of matter, which is also the end product of the alchemical process. He consistently relates the microcosm to the macrocosm, but always speaking imaginatively in *Earth and Reveries of Will*: "Gems are earth stars. Stars are sky diamonds." Certainly, too, he is aligned with alchemy by his singular interest in the four elements, and his evocations of the participatory nature of elements and imagination.

In *Alchemical Psychology*, James Hillman explains what benefit the return to alchemical thinking can have for our time: "Not a literal return to alchemy that is necessary, but a restoration of the alchemical mode of imagining. For in that mode we restore matter to our speech—and that, after all, is our aim: the restoration of imaginative matter, not of literal alchemy."[20] In multiple ways, from radical and unexpected perspectives, Bachelard offers us just this—the restoration of an alchemical mode of imagining.

CHAPTER FIVE

IMAGES AND ARCHETYPES IN BACHELARD AND JUNG

It should not be surprising, then, that to dream material images—yes, simply to dream them—is to invigorate the will.

Depth psychology teaches that earth is archetypal mother, simultaneously preserver and destroyer. Images elaborated by Gaston Bachelard reveal our complex responses to the material world—in will-directed activities and in receptive and restorative movements of psyche. In this chapter we will compare these Bachelardian images with the more symbolic ones of archetypal psychology, particularly those of C.G. Jung.

Let us begin by summoning the early Greek poet Hesiod who gave us some classical images of our "steadfast base of all things," fair Earth. Hesiod, sometimes accused of being a misogynist, here recognizes the essential and indisputable centrality of Gaia—earth, matter, and the common mother of us all.

> Earth, the beautiful, rose up
> Broad-bosomed, she that is the steadfast base
> Of all things. And fair Earth first bore

The starry Heaven, equal to herself,
To cover her on all sides and to be
A home forever of the blessed gods.

What a shock that in our time we are faced, probably for the first time in human recorded time, with grave concerns about the continued health of our planet. In the middle of the twentieth century Albert Schweitzer warned, "Man has lost the capacity to foresee and forestall. He will end by destroying the earth."

I want to explore the archetypal roots of the earth in our psychic make-up and enlarge these with imagery from Bachelard. Though he was a philosopher, many archetypal psychologists have applied Bachelard's original work in images to the field of psychology. It is necessary, though, to clarify some distinctions, especially in methodology, between C.G. Jung and Bachelard. The spectrum of Jung's interest ranges wide in exploring the depths of psychic behavior. As a phenomenologist, Bachelard's search is more limited—concentrated on particular images and image clusters, especially as they appear in verbal communication. He examines the texture and quality of how an image manifests.

Bachelard was at first more influenced by Freud than by Jung, but begins to make reference to Jung in his later books, particularly after 1948. *The Psychoanalysis of Fire* (1938) is certainly a novel approach to a physical element and demonstrates that Freud had caught his attention. In an interview with Aspel (1955) he said that he had "received Jung too late."[1] Bachelard does frequently give offhand compliments to Jung, such as this one about archetypes:

"Everything that originates in us with the clarity of new beginnings is a mad surge of life. The great archetype of incipient life gives to any beginning the psychic energy that Jung acknowledged in every archetype."[2]

Colette Gaudin, in her excellent introduction to *On Poetic Imagination and Reverie,* explains that Bachelard said his lack of medical knowledge prevented him from seeing images as organic impulses. But actually, as she explains:

> The real reason is that he wants to seize the specific originality of the symbol without reducing it to its causes. That is why he favors the Jungian concept of the "archetype," which offers the advantage of including symbolism in the unconscious. Strictly speaking, an archetype is not an image. For Jung, it is psychic energy spontaneously condensing the results of organic and ancestral experiences into images . . . . When Bachelard uses any psychological concept, he limits his investigation to the present life of images; he disregards the historical and anthropological background of archetypes and attempts instead an "archeology of the human soul."[3]

All those who have worked with archetypal images are well aware of the multivalent character of any archetype. This is especially true when considering the archetype of the earth. Earth destroys in order to bring about rebirth— think of the seed that must give way to the flower. Both Jung and Bachelard would agree that the imagination of earth involves us in monumental dimensions, partaking of the very mystery of life—both creation and destruction. As in this poem of Rainer Maria Rilke from *Sonnets to Orpheus XII*, the glorious mystery of creation always startles:

> In spite of all the farmer's work and worry,
> He can't reach down to where the seed is slowly
> Transmuted into summer. The earth bestows.

The earth not only is the source of growth; the earth is equally humus, or the product of decay. These two opposite functions belong together in a syzygy. George Meredith's three lines of poetry, almost like a Japanese haiku, express the paradox: "Earth knows no desolation / She smells regeneration / In the moist breath of decay."

Earth is our intimate home, and language gives away the vibrancy of this relationship. In English, the words "mother," "matter," and "material" are closely associated. The same is true in French and, according to Bachelard, in Hebrew as well. Bachelard frequently uses other writers and poets to prove his points, claiming that "the letter M [the Hebrew Mem], at the beginning of a word, paints all things local and plastic. La Main [Hand], la Matière [Matter], la Mère [Mother], and la Mer [Sea] all would share this initial of plasticity."[4] Jung, in his expansion of the mother archetype, acknowledges the natural affinity of places of protection, saying: "The all-embracing womb of Mother Church is anything but a metaphor and the same is true of Mother Earth, mother nature and 'matter' in general."[5]

Jung has many ways of describing archetypes. Here is one that many will be familiar with: "Archetypes are systems of readiness for action, and at the same time images and emotions." They are "psychic aspects of brain structure," "instinctive adaptation." But here is one that perhaps circulates less widely: archetypes are "that portion through which psyche is attached to nature, or in which its link with the earth and the world appears at its most tangible. The psychic influence of the earth and its laws is seen most clearly in those primordial images."[6] The earth as archetype partakes of the massive energy of any archetype, but, in addition, this particular archetype is strengthened by being the fundamental, the *prima materia*, the basic matrix of all. It is thus doubly dynamic.

Bachelard and Jung agree on the primacy of images. They often sound similar to each other. "Images are primary psychic realities. In experience itself, everything begins with images." This is Bachelard in *On Poetic Imagination and Reverie*, though it could equally well be Jung. Bachelard looks at images as the nucleus of an expanding circle of logical associations. He particularly reveres the literary image. He argues that the language in which we encapsulate our thoughts reveals the essence of each person's individual imagination. Bachelard enjoys classifying poets according to the primary elemental images they most often use. In a poem, we resonate with the poet's imagination and participate in the pleasure of the discovery of new images. Bachelard insists in *Earth and Reveries of Will* that "the function of literature and poetry is to bring new life to language by creating new images." He fully appreciates the immense stimulus of the image in saying: "Indeed, ours is the century of the image. For better or worse we are subjected more than ever before to the power of imagery."

Bachelard turns upside down the simple insistence on the primacy of the real. He affirms "the primitive and psychologically fundamental character of the creative imagination." "In my view," he says in the "Introduction to the Two Volumes," *Earth and Reveries of Will* and *Earth and Reveries of Repose*, "perceived images and created images constitute two very different psychological phenomena requiring a new and special category, to designate the imagined image." Recreated images rely on perception and memory. What he terms "the unreality principle" is equally or more determining than what is normally called "reality":

> Creative imagination functions very differently than imagination relying on the reproduction of past perceptions, because it is governed by an unreality principle .... This unreality principle is every bit as powerful, psychologically speaking, as that reality principle so frequently invoked by psychologists to characterize an individual's adjustment to whatever "reality" enjoys social sanction. It is precisely this unreality principle that reinstates the value of solitude . . . [7]

Surprisingly, it is in our moments of quietude that we experience the rush of images, helping to refine our will to action. Bachelard emphasizes the connection between images experienced in solitude and the will to engage:

> Because reverie is nearly always associated in our minds with a state of relaxation, we fail to appreciate those dreams of focused action which I will call dreams of will. And so, when the real stands before us in all of its terrestrial materiality, we are easily persuaded that the reality principle must usurp the unreality principle, forgetting the unconscious impulses, the oneiric forces

which flow unceasingly through our conscious
life. Only by redoubling our attention then may
we discover the predictive nature of images, the
way that any image may precede perception,
initiating an adventure in perception.[8]

Many may be incensed that Bachelard insists that
the image not only colors perception but also precedes it.
Bachelard sometimes addresses art images, especially in the
short essays in *The Right to Dream*, but for him, dreaming
is daytime reverie. Daydreaming, or reverie, is not simply
wasted time. These are the moments in which imagination
reorganizes the will and translates desire into the will-
to-action. Here Bachelard fundamentally disagrees with
Jung, who seems to denigrate daydreaming as wishful,
unproductive thinking. Further, Bachelard rarely deals with
nocturnal dream images. He doesn't quite trust his ability
to interpret these. This is yet another essential difference
in the way Bachelard works with images, in contrast to
the more symbolic approach taken by Freud and Jung.
Bachelard operates by the logic of linking through sounds
and textures—the way we experience the world through the
five senses. Literary images lend themselves more easily to
expanding in the circular way Bachelard likes to work. His
detractors accuse him of snatching images out of context
whenever he wants to use them to prove a particular point.
But his obvious love of heightened metaphor and created
image makes us forgive him. He accuses psychoanalysis of
not loving images enough:

> Psychoanalysis, like descriptive psychoanalysis
> is reduced to a sort of psychological topology: it

defines levels, layers, associations, complexes,
and symbols . . . . But psychoanalysis has not
developed the resources for a veritable psychic
dynamology, a detailed dynamology which
explores the individuality of images. In other
words, psychoanalysis is content to define
images according to their symbolic meanings.
The instant an impulsional image is divulged or
a traumatic memory uncovered, psychoanalysis
applies a social interpretation. An entire domain
of investigation is neglected: that of imagination
itself. Now, the psyche is animated by a veritable
hunger for images. It craves images. Meanwhile,
psychoanalysis looks for the reality beneath the
image but fails to perform the inverse search:
to look beneath the reality for the absolute
image. It is in this search that one detects the
imaginal energy which is the hallmark of an
active psyche.[9]

Images of air, as an illustration, are difficult to pin
down—they either "evaporate" or "crystallize," Bachelard
remarks in *Air and Dreams*. He shows how the imagination
works by a kind of quick flickering between reality and
daydreaming. Just think of such words as *wing* and *cloud*. We
find ourselves making pictures of them, but then in another
instant the imagination rushes to add new clouds, wings,
and other imaginative thoughts. This happens because the
imagination is naturally dynamic, taking us beyond and
freeing us from only what the senses perceive.

In a chapter on "The Poetics of Wings" in *Air and
Dreams*, Bachelard says: "The sky is a world of wings." He

is such an acute observer of the natural world that is all around, everywhere. Even though it is directly in front of us, how often do we ever stop to really see? He thinks of everything in a new light and with lively connections in such statements as this: "The wing is essentially aerial. We swim in the air, but we do not fly in the water."

Bachelard, our guide to the way images unfold, then goes on with remarks about how birds have always stirred the imagination of all those who love the natural world and the beauty of flight. Religious leaders such as St. Francis, and poets, particularly the English Romantic poets, have sung about this way of raising our thoughts in a heavenly direction. Bachelard describes the joy of some individual birds that have attracted poets: "The startling invisibility of the lark has never been sung better—as a wave of joy—than by Shelley in 'To a Skylark.' Shelley understood that it was a cosmic joy, an 'unbodied joy,' a joy that is always new in its revelation." What this means to me is that it is not only in seeing the beauty of the flight of the bird but even when we only hear its joyful trill that our hearts are lifted. Shelley sees the skylark as a messenger of rapture:

> Teach us, Sprite or Bird,
> What sweet thoughts are thine:
> I have never heard
> Praise of love or wine
> That panted forth a flood of rapture so divine.

Both Bachelard and archetypal psychologists would agree on the innate power of the image, but for Bachelard, "sublimation," a word that often means denial or repression, is a natural function of the psyche:

My position concerning the fundamental nature of the image seems to me true by definition, for I associate the life proper of images with archetypes, the power of which has been demonstrated by psychoanalysis. Imagined images are sublimated archetypes rather than reproductions of reality . . . . What better way to say that the image has a double reality: a psychological and a physical reality. It is through the image that the one who imagines and the thing imagined are most closely united. The human psyche forms itself first and foremost in images.[10]

Bachelard makes no clear distinction between objective and subjective experiences. Person and world are always dialoguing. Bachelard suggests the need for a metaphysics of the imagination:

It seems to me that this is where the elegant work of C.G. Jung himself was leading, when he discovered, for example, the presence of archetypes of the unconscious in alchemical imagery. In a similar vein, I will offer numerous examples of images which are transformed into ideas. Thence we may examine the entire intermediary region of the psyche between unconscious impulse and those earliest images that rise to consciousness.[11]

Jung and Bachelard shared a major interest in alchemical images, although their approach was quite different. Jung delved into the whole dynamic process of the refinement of the base nature of matter into psychic or spiritual essence. Bachelard was more concerned with human interaction with the basic properties of the material world. He staunchly believed in the dynamism of the four classic images—earth, fire, air, and water—those elaborated by medieval alchemists. He devoted one or more books to exploring the depths of these fundamental images of matter. As we have seen, he called them "the hormones of the imagination." Each of these elements has its own special characteristics for engaging our imagination. To oversimplify: fire embodies the most extreme opposites— salvation and apocalypse; water is the most feminine; air desires freedom; earth grounds us all in practical purpose.

It is informative to note Bachelard's reference to Jung and his studies in alchemy in the following statement: "If we read Jung's lengthy study of alchemy, we shall reach a fuller understanding of the dream of the depth of substances. Indeed, Jung has shown that alchemists *project* their own unconscious onto the substances they have long been working on and this unconscious then accompanies sensory knowledge. When alchemists talk about mercury, they are thinking 'externally' of quick-silver while at the same time believing themselves to be in the presence of a spirit that is hidden or imprisoned in matter."[12]

It is not surprising that Bachelard quotes Jung frequently, because in *Earth and Reveries of Repose* he is doing a psychological digging. Along with being a lover of poetry and images, Bachelard is a pioneer, a discoverer of new nuances, an explorer of levels of consciousness. This deep examination is true in all of his works beginning with the publication of *The Psychoanalysis of Fire* in 1938, when he was primarily influenced by

Freud. In *Earth and Reveries of Repose*, the psychological approach is particularly apparent. In the past, Bachelard often criticized psychology for its emphasis on nighttime dreaming while ignoring daytime reveries. Now it is Jung whom he frequently references, explaining that he is doing an *archetypal archeology*. As quoted earlier in this chapter, Bachelard offers this clarification: "An image has a double reality: a psychological reality and a physical reality. It is through the image that the one who imagines and the thing imagined are most closely united."[13] When Bachelard reveals the discovery of an image complex in his readings, he freely refers to it as an archetype, expecting his readers to understand what he means. An image not only reveals hidden traits but, by virtue of its ability to synthesize diverse qualities gathered from various occurrences, it is a force of amalgamation: "An image is, in fact, an integrating force. It converges the most diverse impressions, impressions arising from all the senses."[14]

Part One of *Earth and Reveries of Will* designates complexes of earth after historical or mythological personages, including Atlas and Hercules. Bachelard reminds us of the myth of Atlas, who is said to have held the world up on his mighty shoulders. The hero Hercules, in one of his tests of strength, comes along and offers to share the burden, but gives it back to Atlas. To hold this story in our imaginations for a while causes us to feel stronger ourselves. Our imagination is naturally expansive. When the energy of the imagination is sensed, we begin to feel more and more powerful, and finally all-powerful.

Part Two concerns the deep interiority of those great images of asylum: house, belly, and cave. He chastises the psychoanalyst for a too-facile lumping together of these images, explaining them as an unconscious desire to re-enter the womb: "But this sort of diagnosis ignores the unique values of imagery. There seems good reason to study each of the three avenues of return to the mother separately."[15] The second volume, *Earth and Reveries of Repose,* carries the subtitle *An Essay on the Images of Intimacy.* In this volume, using the same technique, Bachelard employs a dialectical approach in exploring images of external matter as they are experienced in inner life.

Bachelard explains that he had to write two volumes on the images of earth—one representing the extrovert imagination, which "is dedicated to active reveries that invite us to act upon matter." In the second, "the dream flows along a more ordinary incline; it follows the path of involution, carrying us back to our earliest refuges, favoring images of our inmost depths."[16] He explores the Jonah complex, or the desire for interiority. He calls the companion volumes a diptych made up of images of work and images of repose. However, he warns that images of extroversion and introversion only rarely occur in singular isolation:

> In the final analysis all images emerge somewhere on a continuum between these two poles. They exist dialectically, balancing the seductions of the external universe against the certitudes of the inner self. It would be fraudulent then not to acknowledge the double tendency in images to extroversion and to introversion, not to appreciate their ambivalence . . . The loveliest images are often hotbeds of ambivalence.[17]

Bachelard praises the value of work, especially with the hands of the body. Work is most satisfactory when we accomplish it ourselves. Bachelard says this is true in *Earth and Reveries of Will* because "the will to work cannot be delegated. It derives no benefit from the work of others. The will to work prefers to do." Wendell Berry's poem says much the same about the joy of actively working the earth:

> Sowing the seed,
> My hand is one with the earth.
> Wanting the seed to grow,
> My mind is one with the Light.
> Hoeing the crop,
> My hands are one with the rain.
> Having cared for the plants,
> My mind is one with the air.
> Hungry and trusting,
> My mind is one with the earth.
> Eating the fruit,
> My body is one with the earth.[18]

A resistance to human efforts to master or dominate them is a characteristic of each of the elements—earth, air, fire, and water—but the elements other than earth are not *always* hostile. Earth is singular in that respect: "The resistance of terrestrial matter, by contrast, is immediate and constant. From the very first, this resistance becomes the objective and unequivocal partner of our will,"[19] Bachelard explains. In other words, how we manipulate the material world reveals the nature of our will. Our dreams of manipulating matter supply us with a sense of self-definition or unique authentic identity, just as our hands,

87

too, are a principle way we interact with our environment. He adds that "nothing is so unambiguous as matter worked by human hands when it comes to classifying the varieties of human will."

> It is thus that matter reveals to us our own strengths, suggesting a system for categorizing our energies. It provides not only enduring substance to our will, but a system of well-defined temporalities to our patience. In our dreams of matter, we envision an entire future of work; we seek to will. It should not be surprising then that to dream material images—yes, simply to dream them—is to invigorate the will.[20]

For Bachelard, earth is the great schoolhouse, the locus for educating, or leading out the innate but reluctant qualities of human personality: "Matter . . . through the work of the imagination, mirrors our own energies."

Bachelard does not deal with the subtleties of difference between cosmos, earth, and world, or contrarily between earth and underworld. Harkening back to Heidegger's *Origin of a Work of Art*, Ed Casey gives this helpful definition: "Earth is self-including and a historical element while world is communal and linguistic . . . . Earth and world take the measure of things. No wonder they keep on appearing in basic myths, from Plato's *Timaeus* (where chorea and cosmos, primordial place and world, compete constructively) to the Navajo Creation myth . . . "[21]

After three centuries of scientific endeavor to reduce its mysteries, imaginative earth has lost much of its vitality. The myths no longer move us. Primitive peoples were more sensitive to earth's ley lines and magnetic outcroppings, building sanctuaries and temples in locations of high energy. Recent commentators have observed that this knowledge allowed Neolithic cultivators to grow crops in places on sides of hills that no one would dare to use now. In Britain, experienced dowsers have observed the fact that "every megalithic site is over a centre or channel of the terrestrial current whose emanations are detected by the dowser's rod."[22] In China, the art of geomancy—feng-shui ("wind and water")—allowed for a harmony between man and place. The intention and effect of feng-shui was "to produce a landscape which had to provide certain spiritual values and also to fulfill the practical purpose of supporting a dense population."[23] Mircea Eliade complains about our current simplicity: "The completely profane world, the wholly descralized cosmos, is a recent discovery in the history of the human spirit."[24]

Inevitably, we always go back to classical Greek mythology to discern the varieties of little understood aspects of the human, or the superhuman. For our imagination of earth, the image of Gaia gives a sense of cornucopia, of overarching abundance. One of the qualities of Gaia is that of "immovability," according to Pat Berry: "Gaia made things stick. She was the goddess of marriage. One swore oaths by her and they were binding. Mother/ matter as the inert now becomes mother as settler, stabilizer, the binder."[25] Pandora is a creature who provides earth's bounty along with hope to humankind. The myths of Demeter and Persephone relate to the fertility of the earth, bringing into focus the joys of growing wheat. Persephone, queen of the underworld, reminds us of the necessity of the dark underworld to both beginnings and ends of life. The sacred rites of Demeter/ Persephone at Eleusis that lasted thousands of years deny the finality of death and promise the ultimate transformation of life.

The Fates, the Moirai, are involved with earth's terrible power as destroyer of time, cutting off the thread of life and returning us to earth's womb. And of course an entire volume could be written to discuss the fear of the feminine, in its relation to life and death. Perhaps the whole thrust of scientific endeavor is to overcome and control the frightening aspects of Mother Nature and, in the personal arena, the power of our mothers. In contrast, in India the figure of Kali, the Hindu Great Mother, with her foot standing on a pile of bloody heads, is one of the most beloved and respected figures.

Having started this chapter with Hesiod, let us return to him. Since *Theogony*, we know that what first existed was Chaos, or formlessness. From darkness and death, surprisingly, Love was born. And with its birth, order and beauty began to bind up confusion. Love created Light, with its companion, radiant Day. With the coming of Love and Light it seemed natural that the Earth should also appear.[26] It is first love, Eros, and then Gaia who brings form out of confusion and nothingness. Think of that! From the earth come all the starry Heavens. Earth even provides her own escort; she has a hermaphroditic element in addition to her strongly feminine qualities. Pat Berry suggests psychology should not be in such a rush to get rid of turmoil and confusion, for "within chaos there are inherent forms," or "each chaos mothers itself into form." If we look at the image of earth as a picture rather than a narrative, Berry says, "the chaos and forms and earth are given all at once."[26] This means that, as well as being the place of our willful exploits, in a reversal, earth tames us. As Bachelard has said, it mirrors our own energies. Earth is the crucible where our ideas take form.

CHAPTER SIX

THE DUAL IMAGINATION OF EARTH:

Reveries of Will and Reveries of Repose

If we see leaf, flower, and fruit within the bud, this means that we are seeing with the eyes of our imagination.

It is singularly striking that Bachelard, Professor of Philosophy of Science at the Sorbonne, with so many books on science to his credit, can move from a strictly analytical, detached slant on matter to a caring, expansive, encompassing view of the natural world, complete with multiple imaginative associations. As I mentioned previously, Bachelard shared a kinship with Teilhard de Chardin, the metaphysical theologian—to borrow the words of philosopher Christian de Quincey, they both "spoke of the *'within'* of things" (emphasis added).[1]

Bachelard provides this explanation of how imagination and the living world are coupled together, each inspiring the other: "My response would be that the imagination is nothing other than the subject transported into things. Images bear, therefore, the mark of the subject." Thus the images we retain carry our own special stamp. He adds: "And this is so clear a mark that in the end it is by means of images that the most accurate diagnosis of the temperaments can be made."[2] When we get used to the idea of the interconnectedness of the viewer and the viewed, one of the most surprising aspects of this statement is the predictive aspect of material images. Bachelard always claims that, though we may respond to many different elements in nature, we can learn much about ourselves by observing those with which we especially identify.

Bachelard teaches us to be closer observers of the world and also to be more attentive readers. He is not interested in the obvious, the most clearly conscious level of books or the story line, but urges us "to sojourn long enough with novel images, the kind of images that renew archetypes of the unconscious." As mentioned previously, I've often read closely Yeats, and his use of fire imagery, such as these lines from "Sailing to Byzantium": "O sages standing in God's holy fire / As in the gold mosaic of a wall, / Come from the holy fire, perne in a gyre, / And be the singing-masters of my soul."

In his "Preface (For Two Volumes): The Imagination in Matter and in Word" quoted above, Bachelard claims to be devoting his efforts "to defining a beauty intrinsic to material substances, their hidden attractions." He wants to prove that "imagination is not necessarily a vagabond activity, but that, to the contrary, is most powerful when concentrated on a single image."[3] The first study of earthen imagination (*Earth and Reveries of Will*) "marked as it was by the preposition *against* must now be supplemented by a study of images that are marked by the preposition *in*." In other words, the first book is about earth's characteristics that challenge us, while *Earth and Reveries of Repose* directs us inward. That is why the first he called an "extroverted" view of earth and the second, "introverted." *Earth and Reveries of Will* follows the dynamic pathway, the demand to actively engage the will in the world. But even in the earlier book, a whole chapter is devoted to "Soft Matter, The Valorization of Mud." In *Earth and Reveries of Repose*, the more inward approach, looking inside and down deep, he seems particularly involved with created images. Here he is concentrating on images of repose, refuge, and rootedness.

Part of the title in *Earth and Reveries of Repose* actually seems a misnomer. No vital image, as Bachelard always insists, is ever really at rest: "Images of repose and agitation are frequently juxtaposed." By its very nature an active image vibrates with the energy of opposites. Ever oscillating with ambivalence, "ambi-valent" (meaning "many valences"), a vital image has many values on the continuum from positive to negative. That is why it retains the reverberations that keep it active in the imagination: "In the final analysis all images emerge somewhere on a continuum between these two poles. They exist dialectically, balancing the seductions of the external universe against the certitudes of the inner self."[4]

The subtitle, *An Essay on Interiority*, does indicate that he is penetrating into the insides of things. What Bachelard does is shake up our ordinary reactions. He undermines our previous conceptions. He mines the depths of unconscious valuations, taking us into uncharted territory: "*Unnamable* things are retained by the unconscious, they endlessly seek a name." Even the scariest subjects under his scrutiny become avenues of exploration. Many of these subjects are quite creepy, even repulsive. For example, after exploring the secure, stable images of the interiority of the house, he doesn't hesitate when examining other images—of belly, roots, caves, serpents, labyrinths—to reveal the unseemly side of material things. For instance, the serpent "is one of the most important archetypes of the human soul. It is the most *earthen* of animals. It is truly an animalized root and, where images are concerned, the link between the vegetable and animal kingdoms."[5] A few paragraphs later, he explains our intense reaction to serpents: "Emotion—that archaism—governs the wisest of people. When we are faced with a serpent, a long line of ancestors is there in our troubled soul, all of them filled with fear."[6] Soon after, he adds this thought: "The serpent is thus quite naturally a *complex image* or, more precisely, a *complex of the imagination*. It is imagined as giving both life and death, as supple and hard, straight and round, motionless or swift moving. This is why it plays such a part in the literary imagination."[7] Bachelard can find beauty in the ugly, or terror in the beautiful, reminding me of Coleridge, whose mariner in "The Rhyme of the Ancient Mariner" begins to find redemption for having killed the albatross by finding beauty in the slimy creatures under water in the ocean deep.

Bachelard approaches the world with an attitude of wonder. He values direct experience of the material world because the world beckons us to participate. It is hard to decide which response he prefers—the earth that lures us with a call to action, or the after-reflection that provides endless metaphors in which the role of imagination is more obviously entailed. Bachelard apologizes for not discussing agriculture in either book on the earth, which he says, in *Earth and Reveries of Will*, would take a whole book in itself: "It is not due to any lack of the love of the soil," adding this explanation: "It would require an entire book to unearth the agriculture of the imagination, the joys of spade and rake." He praises the value of physical work when we engage the hands or the body. Work is most satisfactory when we accomplish it ourselves instead of hiring someone else. Bachelard says this is true because "the will to work cannot be delegated. It derives no benefit from the work of others. The will to work prefers to do."

The imagination is always engaged in any satisfactory endeavor, because the imagination, for Bachelard, is primary, and precedes any action or observation. Even the simplest action of walking from here to there, we can only do what we can imagine doing. Imagination then takes experience and enlarges upon it, searing it in our souls. In the spoken or written image we encounter further resonances. In this book, the last that he finished writing on the four elements in 1944, Bachelard examines after-images or created images: "Creative imagination is different than reiterative imagination which is ascribed to perception and memory."

We are constantly changing images into ideas, ideas into images. Bachelard states his case, in *Earth and Reveries of Repose,* for the exchange of image and idea primarily by examination of the literary image: "Thus, the *literary image* is privileged in that it acts as both an image and an idea. It can imply something interior to us and also something objective." A literary image is no mere copy of the exterior world. Bachelard turns around conventional thought: that we first view the natural world and afterwards reproduce a duplicate in imagination; instead, in *Earth and Reveries of Will*, he claims that "Imagined images are sublimated archetypes rather than reproductions of reality." This thought is really quite radical. When he says that "the literary imagination has a subtextuality of its own," he means that the literary image calls forth ripples of nuances and associations all its own, providing the metaphors used to express any particular experience. He stresses the difference between perception and the later use of an image by a poet, with convincing proof that "literary expression enjoys an existence independent of perception, that the literary imagination is not a secondary form which can only succeed visual images registered by perception."[8] These links are available to all if we allow ourselves the time to be open to reveries and to let associations flow that give depth to any experience. Instead of racing along with only superficial involvement, the careful reader gathers layers of associations that make life instantly richer and more meaningful.

In *Earth and Reveries of Repose*, Bachelard says that the poet can misunderstand botany and still write a beautiful line: "If we see leaf, flower and fruit within the bud, this means that we are seeing with the eyes of our imagination." In Yeats's poem "Among School Children," the poet juxtaposes the stages of man from birth to death with the aging of a mighty tree. The last stanza carries the image of the chestnut tree in all three stages of development. The narrator of the poem asks: "O chestnut-tree, great-rooted blossomer, / Are you the leaf, the blossom or the bole?"

Bachelard is forever creating and recreating his

dynamic definition of imagination, which for him is the most primary of all human functions. In *Water and Dreams: An Essay on the Imagination of Matter* (1942), he stated his case: "The *imagination* is not, as its etymology suggests, the faculty for forming images of reality; it is the faculty for forming images which go beyond reality, which *sing* reality." In *Air and Dreams: An Essay on the Imagination of Movement* (1943), he developed this further: "The value of an image is measured by its *imaginative* aura. Thanks to the *imaginary*, imagination is essentially *open* and *elusive*. It is the human psyche's experience of *openness* and *novelty*. More than any other power, it is what distinguishes the human psyche. As William Blake puts it: 'The Imagination is not a State; it is Human Existence itself.'"[9]

In the 1960s, Bachelard was still enlarging his comprehensive view of the imagination as not just a primary quality, but an essential quality of human consciousness. Shortly before his death in 1962, in the unfinished book *Fragments of a Poetics of Fire*, he lamented his inability to fully define the extensive power of imagination. Even though his work on images had already filled over 2,000 pages, he revealed this wistful thought: "I wish I had all of my books to write over. It seems to me that I should know better now how to express the reverberation of spoken images in the depths of the speaking soul, better now how to describe the links between new images and those with ancient roots in the human psyche."[10]

In *Earth and Reveries of Repose*, Bachelard once more stresses "imagination's all-consuming *desire* to go deep into matter." He again verifies the singularity of imagination: "The imagination is more interested in the newness reality offers, in what matter reveals. It loves the open materialism that unceasingly presents itself as opportunities for new, deep images. In its own way, the imagination is objective."[11] What keeps a sight we have often experienced before, such as a glowing sunset, ever thrilling, is its imaginative associations. In addition, ultimately it is the similarities in the extent and power of imagination that link us to our fellow humans and to those who have lived before. Bachelard explains this connection: "A certain homogeneity in human imagination spans the centuries—proof for me that the imagination inheres in human nature itself."[12] The imagination is not an exclusive characteristic of the artist or poet; the imagination is essentially vital in everyone.

We are renewed when we follow Bachelard's instruction on how to read the world or how to read a poem. For all his inspiriting of matter, his enlivening of the world, there a remains always a down-to-earth cast to his thoughts. Bachelard's contribution to the understanding of the imagination, indeed to the whole of consciousness, is measureless.

CHAPTER SEVEN

THE HAND OF WORK AND PLAY

Work is its own therapy, with benefits that carry deep into the life of the unconscious.

When I first read Bachelard's *The Poetics of Space* as a graduate student in the Psychology department at the University of Dallas, it stirred my imagination in an entirely new way. What we were studying was a specific form of psychology/philosophy called "phenomenology." It meant, at least on one level, that we examined with fresh attention how each of us individually respond to the physical world. Before making generalizations about such complex reactions as anger or even joy, or before describing the physiological manifestations of an emotion, such as the surging blood in the face of anger or shame, we asked ourselves how we experienced each of these moments subjectively. Collette Gaudin, in her introduction to Bachelard's *On Poetic Imagination and Reverie*, recognizes the impact of phenomenology on his thought: "In a word, the phenomenological approach is a description of the immediate relationship of phenomena with a particular consciousness; it allows Bachelard to renew his warnings against the temptation to study images as things. Images are 'lived,' 'experienced,' 're-imagined' in an act of consciousness, which restores at once their timelessness and their newness."[1] This Bachalardian approach bestows on images a livingness value that helps to explain their power to engage us.

I remember one day being somewhat stunned when asked by a professor a question about the common coffee cup—what it implied about being human. After much pondering, I came up with a few scant observations, such as, a cup approximates the proportions that a human being can comfortably wrap a hand around. Directing my attention in this way to function and form started my thinking about the relevancy and significance of the human hand to the lived world.

The two most expressive features of the human body are probably the eyes and the hands. We are told that the soul is revealed through the expression of the eyes. Think of Beatrice leading the pilgrim Dante upward, traversing the rings of Paradise with her ever-brightening eyes in *The Divine Comedy*. Next, in revealing the essence of being human, is surely the hand.

Consider for a minute a few of the multiple ways we use our hands to declare ourselves. Think of the Buddha. The extended right hand of the Buddha has a name, *semui-in mudra*, which "has the power of giving tranquility and grants the absence of fear." Buddhism has a whole science of hand positions, with a psycho-symbolism of each finger. We might call it sexist, but the index finger, pointing, is considered masculine, independent, and the next finger is feminine, patient. Think of the Roman Emperor ("Hail

Caesar") or Hitler ("Heil Hitler"). Many other examples of the expressiveness of the human hand in art history come to mind—I can't help remembering the Sistine Chapel, with God reaching out His hand to convey the spark of life to Adam through an outstretched finger. The hand of Christ on the beautiful altarpiece at Christ the King Catholic Church in Dallas has the first two fingers upright pointing upward. Maybe I'm becoming a hand fetishist—I see hands everywhere.

The hand is an avenue of communication, independent of the verbal medium. A baby will point and put its hands over eyes or ears in pre-verbal expression, or clap its hands to convey joy. Have you ever played the hand game of "Here's the Church and Here's the Steeple"? Or made a Cat's Cradle? Or the one where you stretch a string and make a cradle's bow? What we do with our hands can define a complete attitude or our whole relationship to others. Think of the thumbs-up or thumbs-down gesture. Having someone rudely flash us the finger is a gesture that is a too- frequent occurrence on the road.

In the Eastern tradition, hands are folded together to acknowledge a respectful attitude toward the other in meeting. When we finish practicing yoga, we bring our hands together and say our thanks, "Namaste." In the West, we greet each other or make a deal by shaking hands. We use the hands to say "Goodbye." We raise our hands to say we want to speak. We put our hand on the Bible to make an oath. We might say in a serious moment: "It's out of my hands; it's in the hands of God." We pray by clasping our hands together. We make the sign of the cross by passing our hands from left to right and up to down over our heart.

The Hand of Work and Play

In the Christian litany, Christ sits at the right hand of God, just as we honor a guest by choosing a place on our right hand. For left-handedness, the word in French is "gauche," indicating a difference from the usual right.

Both science and studies of the body emphasize the hand. For the deaf, sign language is a powerful use of the hands as a tool of communication. If you have ever tried to read Braille for the blind, you will realize how much sensitivity the fingers can develop. A recent study reported in *Nature* indicates that gesturing with the hand helps people lay out abstract thoughts or recall words. Gesturing isn't learned. Children blind from birth use hand gestures, even when speaking to other blind children. In this day of advanced electronics, transactions on the New York Stock Exchange are still conducted with hand signals. When we hurt, we instinctively put our hands on the spot. The healer too uses the hand to massage or stroke the body. It is through the hands that we are most aware of touching and being touched. Rubbing the hands together generates recuperative heat. We do it to warm ourselves. In experiments at the University of Texas Southwestern Medical School, many effects of biofeedback can be graphically demonstrated. Through mental concentration with directed attention on the hands, not actually rubbing them, their physical temperature can be raised. The fortune teller looks at our hands to discern our length of life, our talents, and our destiny.

Whether in measurement or play, the hand is a direct means of communication. We say something is "close at hand." Or we want a handful of berries or a pinch of salt. A horse's height is measured in hands (four inches). The clock has hands. We employ many expressions about hand use. We have time on our hands. We take things in hand, show our hand, tip our hand, or throw up our hands. Hand words invade our vocabulary. We talk of keeping a hand in a project. "Many hands make light work." "Idle hands are the devil's." Things are handmade, or letters hand-written. We live hand-to-mouth or receive handouts or hand-me-downs; we can be even-handed, short-handed, underhanded, heavy-handed. We say, "I've got to hand it to you." The exclamatories "Hands Off!" or "Hands Up!" get the point across quickly.

Hands are multifunctional. Joining hands before saying grace or at the beginning of a class represents coming together in a common endeavor, cooperation, solidarity. Many games such as tennis and ping-pong require hand-eye coordination. Most musical instruments require trained use of the hands. Lovers hold hands. The Beatles made famous the song "I Want to Hold Your Hand."

Frank R. Wilson, in his book *The Hand*, gives some stimulating information about the anthropological development of the physical hand. He sees the hand as the key to the evolution of intelligence—the reason humans developed such large brains lies in the peculiarities of the human hand. Our prehensile thumb allowed us to wrap our hand around a stone and make it a weapon or a tool. Without our fancy mitts, he argues, there wouldn't have been any point in growing big brains.

By profession, Wilson is a neurologist and interna-tionally recognized specialist on the relationship of hand use to human cognitive and artistic development. He is also the former medical director of the Peter F. Oswald

Health Program for Performing Artists at the University of California School of Medicine in San Francisco. He works with musicians who have manual disorders, and his book celebrates the importance of hands to our lives today as well as to the history of the species. In a *New York Times* review of Wilson's book, David Papineau explains:

> The current favorite theory of why we grew large brains that distinguished us from our hominid ancestors is that the purpose was to gossip. Less facetiously, the idea is that hominids with enough brainpower to be interested in the psychology of others had a great advantage once social cooperation became the norm. Another rather older answer to the evolutionary conundrum is that the biped stance we adopted when we left the trees made for a wide range of flexible behaviors. Wilson accepts that both of these adaptations played a part. But he insists that neither would have made any difference were it not for the special structure of our hands.[2]

For Bachelard, the hand defines the human being as no other physical part of the body. "The hand as well as the eye has its reveries and poetry," he says. In *Earth and Reveries of Will* we again draw on Bachelard's particular definition of reverie, which he regards as far more vital to imagination's activity or to psychic balancing than the nocturnal dream. " . . . [A] daydream . . . is less *insistent* than a night-dream," he insists in *Poetics of Space*. The sense of the "I" is never lost. Reverie for me is that time in the morning when I don't jump immediately out of bed.

I can't remain awake long enough at night. When half-dozing, if I can refrain from making lists of things that I need to accomplish, I try to be an observer of the images that flow through my semi-conscious state. Others have somehow made day-dreaming sound like a waste of time. Not Bachelard. He reflects in *Poetics of Reverie* that all of our good ideas and plans begin in images formed when we are quiet and receptive. Reverie is a playful time for Bachelard when we can approximate the state of wonder that we experienced in childhood. Such moments are very restorative to the psyche, because we are momentarily "liberated from the gear-wheels of the calendar."

In exquisite detail Bachelard explains how we fall asleep: "After the relaxation of the eyes comes the relaxation of the hands, for they too come to reject objects." The hands, those seekers of work during the day, relax at night. Indeed, relaxation of the hand is a requisite if we wish to sleep: "When we bear in mind that the whole specific dynamism of the human being is *digital*, it follows inescapably that oneiric space unfolds as and when our knotted fists unclench themselves."[3] Digital here does not refer to anything mathematical or technological.

One of the aspects of Bachelard that I most appreciate is his acceptance of human anger not as a liability but often as a spur to action. Tantalizingly, he says in *Earth and Reveries of Will*: "To imagine a fist clenched for no reason would be a dishonor to the high drama of anger and mar the image of invincibility." Anger is actually energizing, helping to focus attention.

Bachelard honors the entire spectrum of what it means to be a human, foibles and all. Bachelard makes us aware of nuances of soul that we have always taken for granted and persuades us to look at the world anew, whenever possible, with the openness to surprise and wonder in the eyes of childhood.

Bachelard has an almost reverential attitude toward the imagination. He considers the imagination not only the source of pleasure and satisfaction but, more important, the primary source of stirring, vitalizing, and galvanizing our actions. Often the mind, which is needed for accomplishing goals, and imagination are at odds with each other. Satisfying the mind so often means doing violence to the imagination. He urges us to give imagination full play, to allow ourselves to enjoy the jolt of joy that imagination stimulates before plotting how to effectuate any plans. We are most happily productive when physical action, work, and the images of reverie coincide. Then we can mold the world to our inner model. We can get a grip on it, to reemphasize our ever-present hand.

For Bachelard, imagination is the galvanizer of will and supplier of energy to take action: "Where the imagination is concerned, one is only strong if one is all-powerful. Reveries of the will to power are reveries of the will to be all-powerful." Bachelard illustrates this point by referring to Greek mythology: "If one could study the labors of Hercules as dynamic reveries, as images of the primordial will, one might gain access to a *central hygiene* almost as effective as *actual hygiene*. To imagine effort lyrically, to endow efforts conjured in imagination with all the splendor of legendary imagery, is truly to condition one's entire being and to do so without privileging some muscles over others as is usual in purely physical exercise."[4]

The possibilities of what the hand can accomplish prompt us to engage in activities of the world and to dream that we can conquer whatever is necessary. The capable hand gives us confidence that we can overcome any difficulties. We think, "I can hand-le that." To handle something is to get control over it. The world presents jobs waiting to be done. The destiny of work is present in our bodies. In *The Right to Dream*, Bachelard alludes to the clenched fist, which is linked to toil and needs toil, as " . . . the digital will . . . a will to build." Being endowed with a hand, we can dream of holding the world in it. There is a song about God having the whole world in his hand, isn't there?

Dreams of power begin to gather form in the imagination. These reveries move toward the strengthening of the personality or toward aggressive action in the world. The power of the will is first brought into play when faced with matter to be "taken in hand." We can use the hand to fashion a tool to weed the garden, or a tool to strike the enemy. In our prehistoric past, Bachelard traces the move from weapon to tool and claims that at that juncture humankind's moral nature was born: "Humanity's greatest moral triumph came with the invention of the work hammer by means of which destructive violence was transformed into creative power."[5] True to the power of the imagination, what the hand can use for defense can also be used for eliminating some of the toil from work. In Bachelard's words:

> Hammer and club embody the opposition
> of good and evil . . . At first, the simple stone
> clutched in a fist enhanced human cruelty,

forming the first weapon, the first mace. Adding a shaft to the stone served only to extend the arm's own violence, the stone a sort of fist at the end of the forearm.

But the day came when the stone hammer was used to cut other stones, and *indirect* thoughts, prolonged indirect thoughts, stirred to life in the human brain, intelligence and courage formulating together an application of energy. Work—work against matter—is the immediate benefit . . . . With the hammer was born an art of *short, sharp blows of force rapidly delivered,* a consciousness of exact will.[6]

Bachelard extols the benefits of work to liberate restrictions or limitations, what we might call the hang-ups in the psyche. He recommends the efficacy of "striking blows against troubles and annihilating them." And in addition: "Work is its own therapy, with benefits that carry deep into the life of the unconscious[7] . . . . The worker is [transported] into regions where will is liberated from the phantasms of primal impulse."[8]

It is Bachelard's reverence for matter that distinguishes his approach, his unique methodology. The attention he calls to the physical world causes us to observe it more keenly, to heed it more acutely, and to honor it more fully. He contradicts the scientific view that often dismisses matter as inanimate, an attitude more prevalent in the first half of the twentieth century. He insists that matter is totally alive.

Bachelard, who enjoys general renown in France,

is most familiar to literary critics in the United States. In writing a preface to *Earth and Reveries of Will*, I wondered why environmentalists haven't latched onto Bachelard. He is not interested in large-scale projects to save the planet, but his regard for material substances inspires one to care for the land underfoot. Metaphysically we might say he is concerned with the energizing and *in-spiriting* of matter. His earliest foray into the studies of imagination involved analyzing, even psychoanalyzing the four classical material elements—earth, air, fire, and water—his "hormones of the imagination." "Material elements reflect our souls; more than forms, they fix the unconscious, they provide us with a sort of direct reading of our destiny."[9] In describing the Bachaelardian approach, Collete Gaudin says in the book cited above: "The audacious idea of psychoanalyzing elements drew immediate attention to the originality of Bachelard's research."[10]

In his first book on earth, *Earth and Reveries of Will*, Bachelard takes up the images of various material substances that challenge us to engage with the world. "Rocks teach us the language of hardness." Rocks or any hard substance "serve as educators of the will." Bachelard quotes Gerhardt Hauptmann, a German poet, to accentuate his point that the hand is more than a mere appendage: the hand wants to accomplish tasks:

> I feel it in my arm – 'tis hard as steel;
> And in my hand, that, as the eagle's claw,
> Clutches at empty air and shuts again,
> Wild with impatience to achieve great deeds.[11]

Here is an example of the literary use of the image of

the hand from Gabriel Audisio, one that emphasizes the strength of the clenched fist: "Thank you for your lessons in hatred and vengeance / I'll make myself harder than your clenched fists."[12]

At the opposite end of the scale of hardness from rocks, Bachelard calls attention to the image of *pate*, to which he gives a very broad meaning. *Pate* represents "any malleable matter (dough, clay, molten metal); it acquires the value of a central and direct metaphor of imaginary life."[13] "*Pate* is not only an example of material combination (earth and water), it also provides the paradigms of reverie and spiritual life." Imagination always moves toward manifesting its images gathered in dream or reverie: "The space in which the dreamer is immersed is a 'plastic mediator' between man and the universe," he claims. "While provoking our creative energy [*pate* or any of these mediums] also stimulates our consciousness and thus gives rise to an intense happiness."[14] These malleable substances offer the physicality of a hand the glorious opportunity to mold matter to human dimensions.

For Bachelard, it is true that the trusty hand provides happiness by "linking liberation of the soul to work." He devotes many passages to the unique joy experienced by the hands engaged in working clay or kneading dough. This creative endeavor also participates in the pleasure of combining two elements—water and earth. The tactile experience (sometimes done with closed eyes) stirs dreams: "The hand also has its dreams and its hypotheses. It helps us to understand matter in its inmost being. Therefore it helps us to dream of it."[15] Bachelard draws many examples of the possible levels of wetness of dough or paste and the hand's enjoyment of fashioning a vessel or, with flour, a

111

cake. But it is not only the hand that finds satisfaction in mud. When the weather is warm, we are always tempted to take off our shoes and feel the sand or grassy earth against our feet. Bachelard writes about this experience: "Walking barefoot in . . . mud awakens our own primitive, natural connections with earth."[16]

Bachelard is sometimes criticized for using poetic images as examples to prove his thoughts. Generally, he disregards the whole poem but just lifts pieces to place emphases or underline an image. But perhaps he can be excused for this tendency because he has the greatest regard for the poet and for poetry as a conveyor of life's essential meaning: "The poet helps us to discover destinal forces." In other words, the poet leads us in a teleological way to follow our unique star. The poem opens up a special place for us, one where we are enraptured by being temporarily out of causal time: "Everything, in short, that loosens the ties of causation and reward, everything that denies our private history and even desire itself, everything that devalues both past and future is contained in the poetic moment."[17] And further, "Poetry is a metaphysics of the moment. . . . It does, however, demand a prelude of silence. . . . We have here neither the spirited, masculine time that thrusts forward and overcomes, nor the gentle submissive time that weeps and regrets, but the androgynous moment. The mystery of poetry is androgynous."[18]

When I think of hands, I think of the amusing turn that Yeats took in his poem "Broken Dreams." His life-long love for the beauty of Maude Gonne was legendary. He often compared her to Helen of Troy. In another place he spoke of her as "the woman Homer sung." But here he remembers her hands as her one flaw:

> You are more beautiful than any one,
> And yet your body had a flaw;
> Your small hands were not beautiful,
> And I am afraid that you will run
> And paddle to the wrist
> In that mysterious, always brimming lake
> Where those that have obeyed the holy law
> Paddle and are perfect. Leave unchanged
> The hands that I have kissed,
> For old sake's sake.[19]

It may seem that I have favored the image of the hand at work more than at play, but Bachelard would insist on the interchangeability of the two engagements—that work is really play when imagination is fully engaged. Finally he insists in *The Right to Dream*: "Every hand is an awareness of action."

CHAPTER EIGHT

THE FULLNESS OF SILENCE

Created in the silence and solitude of being, with no connection to hearing or sight, poetry seems to me to be the primary phenomenon of the human aesthetic will.

With their boulders and patterned stone, Zen gardens capture the sense of quietness within the ever-moving path of life. If we are able to give ourselves to the experience, we feel our diaphragms relaxing and our breathing lengthening out. This same sense of sanctuary is also available when we are able to internalize moments of stillness. The silent sanctuary that we can learn to tap can remove us from the daily "stings and arrows of outrageous fortune."

In "The Second Coming," Yeats gives us a frightening image of Western civilization, an image of an expanding vortex where "things fall apart" and "the falcon cannot hear the falconer." In this dizzying confusion, "the centre cannot hold," as rationality is spinning out of control. At that particular time, Yeats was addressing a world heading toward the conflagration of World War II. But in our manic, electronic age, much of the disorientation he envisioned in 1938 occurs daily as we rush around at warp speed.

Although we can't stop the world and get off, aren't there benefits to the daily practice of slowing our thoughts and entering in a place of quietude and solitariness? Many are discovering that a short time for meditation, ten or fifteen minutes, not an easy practice in our hectic days, can calm the roving spirit and pull us back to that inner place where, instead of falling apart, we find our focus. The great Sufi poet Rumi in the 13th century wrote, "Only let the moving waters calm down, and the sun and moon will be reflected on the surface of your Being." In other words, when the agitated waters of thought and emotion still, the glory of singular Being can shine forth.

The necessity of entering into silence has been acknowledged by all the world's great spiritual leaders. The *Tao Te Ching* supports this reassuring thought—that concentrating on the root of quietness will prepare us even for the transition through death in a state that will continue to exist even after the body dies:

> Push far towards the Void,
> Hold fast enough to Quietness.

> This return to the root is called Quietness,
> Tao is forever and who possesses it,
> Though his body ceases, is not destroyed.[1]

Even though it may be hard to accept that returning to the root of quietness prepares us for the transition of death, solitude is truly the source of solace. We find our lost selves in silence. Then why is it so often shunned or actively feared? Why do we fill every minute with excessive activity, trying our best to avoid solitude? Perhaps there is a premonition of the silence of death that is frightening. Or perhaps we prefer distraction to the hard work of delving deeply to put together the assorted pieces of our personality.

In 1948 Max Picard, the Swiss philosopher, wrote that very satisfying book *The World of Silence*. At the time we were entering what has been called the post-modern world, but before the impact of the information age had really taken hold. Already appreciating the fast-disappearing value of silence, he explains:

> Man is better able to endure things hostile to his own nature, things that use him up, if he has the silent substance within. That is why the peoples of the East, who are still filled with the substance of silence, endure life with machines better than the peoples of the West, whose silent substance has been almost completely destroyed. Technics in itself, life with machines, is not injurious unless the protective substance of silence is absent.[2]

Silence also has a bearing on our perceptions of time. I have become convinced of how rewarding it is to expand our subjective sense of time by periodically entering into the inner world of silence. A virginal world, silence is full of all possibilities and at the same time, complete in itself. Picard echoes this sentiment: "Silence contains everything within itself. It is not waiting for anything; it is always wholly present in itself and it completely fills out the space in which it appears. It does not develop or increase in time, but time increases in silence." The reason for the sense that in silence, time is elongated is that "Past, present, future, are in a unity in silence."[3] Lovers entering together into a joint world of shared silence have a heightened sense of clairvoyance or premonition due to the supra-temporal quality of this inner space.

Gaston Bachelard brings characteristically unusual thoughts regarding the underappreciated values of silence. He calls silence "reverie," because for him, reverie is an active fullness, not an emptiness. In moments of reverie we discover the integrating force that brings all the ravaged, opposing pieces of the personality together and unites the subjective self with the objective world: "Reverie reconciles the world and subject, present and past, solitude and communication." Supporting Bachelard in his insistence on the efficacy of silence, Picard emphasizes the positive qualities:

> When the substance of silence is present in a man, all his qualities are centred in it; they are all connected primarily with the silence and only secondarily with each other. Therefore it is not so easy for the defect of one quality to infect all the others, since it is kept in its place by the silence. But if there is no silence, a man can be totally infected by a single defect so that he ceases to be a man and becomes so entirely identified with the defective quality that it is as though the defect and the evil it represents were covered merely by a human mask.[4]

Silence provides an opening into psychic space where we regain a sense of wholeness. Each moment we linger in silent attentiveness is a new window of opportunity where time makes a fresh start. Picard explains this point clearly:

> In silence, therefore, man stands confronted
> once again by the original beginning of all
> things: everything can begin again, everything
> can be re-created. In every moment of time,
> man through silence can be with the origins of
> all things. Allied with silence, man participated
> not only in the original substance of silence but
> in the original substance of all things. Silence
> is the only basic phenomenon that is always at
> man's disposal. No other basic phenomenon is
> so present in every moment as silence.[5]

For Bachelard, silence is vital not only as restorative, but as absolutely necessary for defining the authentic and singularly creative in an individual. "The passions simmer and resimmer in solitude: the passionate being prepares his explosions and his exploit in this solitude."[6] It is in our moments of quietude that we experience the rush of images, those images that help to define our very being by galvanizing the will to action. Reverie is not merely for recuperation or rest. Solitude provides the active matrix for will and image to amalgamate and then to seek expression. Bachelard recasts the definition of reverie:

> Because reverie is nearly always associated in
> our minds with a state of relaxation, we fail to
> appreciate those dreams of focused action which
> I will call dreams of will. And so, when the real
> stands before us in all of its terrestrial materiality,
> we are easily persuaded that the reality principle
> must usurp the unreality principle, forgetting

the unconscious impulses, the oneiric forces which flow unceasingly through our conscious life. Only by redoubling our attention then may we discover the predictive nature of images, the way that any image may precede perception, initiating an adventure in perception.[7]

What is unusual is that Bachelard insists that it is most natural for reverie to seek written expression in poetry. In a wonderful chapter called "Silent Speech," he explains that unuttered vocalization becomes poetry when "reflective silence" is joined to "attentive silence." Poetry, created out of the "silence and solitude of being," reveals that will and not emotion is the first impulse of poetry:

> According to the principle of projection, the word is willed before it is spoken. In this way, pure poetry is formed in the realm of the will before appearing on the emotional level. For this reader it is all the more true that pure poetry is far from being the art of representation. Created in the silence and solitude of being, with no connection to hearing or sight, poetry seems to me to be the primary phenomenon of the human aesthetic will.

> Willed and re-willed, the origins of poetry's vocal values are cherished in their essential expressions of will. As they join together, these values give rise to symphonies in our nervous system that animate even the most silent being. These are the most lively and playful of all dynamic values. The will finds them in silence and emptiness of being, at a time when it does not have to set our

119

muscles in motion, but rather when it indulges
in the irrationality of an innocent word . . . .
Before any act, we need to say to ourselves, in
the silence of our own being, what it is we *will*
to become; we need to *convince ourselves* of our
own becoming and to *exalt* it for ourselves.
This is the function that poetry plays in
questions of will. The *poetry of will* must then
be put in touch with the tenacity and courage
of a silent being.[8]

Bachelard considers the writing of poetry a highly
creative act for which silence is the first requisite. The
point is not just to hear our own thoughts, but to hear the
universe speaking. By inversion, "The world imagines itself
in human reverie." Here in this passage he elaborates how
poetry heals the ravages of time:

There are also poets of silence who start by
shutting off the clamor of the universe and the
roar of its thunder. They hear what they write
as they write, in the slow measure of a written
language. They do not transcribe their poetry,
they write it . . . Through the slow rhythm
of written poetry, verbs recover their precise
original movements. Each verb is re-endowed,
no longer with the time of its utterance, but
with the true time of its action. Verbs that spin
and those that shoot can no longer be confused
with each other in their movements. And
when an adjective gives flower to its substance,
written poetry and literary image let us slowly
experience the time of its blossoming. Poetry

then is truly the first manifestation of silence.
It lets the attentive silence, beneath the images,
remain alive. It builds the poem on silent time, a
time upon which no rhythmic beat, no hastened
tempo, no order is imposed. It builds on a time
open to all kinds of spirituality and consonant
with our spiritual freedom.[9]

A poem, born in reverie, releases the freedom of
words. Summing up Bachelard's philosophy, Gaudin writes,
"Reverie shatters frozen meanings and restores to old words
ambivalence and freedom." He never believes that manual
or mental work in the material world provides merely an
experience of reality. Of far more importance for the flow
of creative imagination is what he calls "the unreality
principle," only accessible by entering into silence:

Creative imagination functions very differently
than imagination relying on the reproduction of
past perceptions, because it is governed by an
unreality principle . . . . This unreality principle is
every bit as powerful, psychologically speaking,
as that reality principle so frequently invoked
by psychologists to characterize an individual's
adjustment to whatever "reality" enjoys social
sanction. It is precisely this unreality principle
that reinstates the value of solitude.[10]

We cannot feel if we are too full of perceptions.
Bachelard would argue that we fit our perceptions to our
inner images. In a metaphysical vein, Bachelard states in *The
Poetics of Reverie* that reverie unites us with others in an act
of love: "Reverie is an active force in the destiny of persons

who wish to unite their life through a growing love." Picard is even more explicit:

> Silence is the firstborn of the basic phenomena. It envelops the other basic phenomena—love, loyalty, and death; and there is more silence than speech in them, more of the invisible than the visible. There is also more silence in one person than can be used in a single human life. That is why every human utterance is surrounded by a mystery. The silence in a man stretches out beyond the single human life. In this silence man is connected with past and future generations.[11]

In my moments of silence I feel very much in touch with my father who was killed over fifty years ago in World War II. I often hear my father's and my mother's words. In silence, I feel unseparated from my husband, who died thirteen years ago. I feel that I can somehow share in the process they are going through. Again Picard captures the essence of the experience:

> Silence gives to things inside it something of the power of its own autonomous being. The autonomous being in things is strengthened in silence. That which is developable and exploitable in things vanishes when they are in silence. Through this power of autonomous being, silence points to a state where only being is valid: the state of the Divine. The mark of the Divine in things is preserved by their connection with the world of silence.[12]

The Fullness of Silence

In considering the fullness of silence, I keep thinking of images that gather us up in their silence. I am reminded of works of art that, without words, seem to express the heart of silence. Images, the building blocks of dreams, poetry, and art works, are silent speech. Picard explains:

> Images are silent, but they speak in silence. They are a silent language. They are a station on the way from silence to language. They stand on the frontier where silence and language face each other closer than anywhere else, but the tension between them is resolved by beauty . . . . It is the soul that preserves the silent images of things. The soul does not, like the mind, express itself about things through the medium of words, but rather through the images of things. Things have a dual existence in man: first in the soul through images, then in the mind through words.[13]

Gordon Parks, a renowned photographer and great friend, in his book *Arias in Silence,* gives an image to the heart of silence, and to many of the ideas I have been trying to describe. Here is the poem from which the book takes its title:

> Encamped in the dream
> I felt the moon touch my bed
> It smiled, lingered for a moment,
> left me with a sense of something ahead.
> I tried stitching those voices together,
> but they hung mute in shadows.
> Lately they repose in me
> like a gray remembrance,
> leave me with dreams so brittle to touch.

Now, when every dawn arises,
I still hear those voices singing,
even roaring at times—in silence.[14]

Is there a time when silence is not a positive place of refuge and renewal? When is it evidence of isolation and loneliness, as in some of the paintings of Edward Hopper or Andrew Wyeth? When is silence a painful, suffering *Silencing of the Self*, the name of Dana Crowley Jack's book? The silence we are celebrating doesn't mean restrictions on self-expression or "freeze-drying" the emotions. All of us have experienced moments when we should have spoken but didn't, for fear of sounding foolish or revealing something about ourselves that we preferred to hide.

Dana Crowley Jack explains that keeping silence instead of expressing our thoughts can lead to depression. Both women and men can deny some part of essential being, but Jack feels that women are particularly prone to this self-abnegating behavior which leads to an accumulation of resentment and anger: "Depressed women's self-silencing appears to be a more extreme form of ordinary self-censoring."[15] In addition, squashing our thoughts carries unexpected consequences—in a relationship, it reinforces hierarchy, expressing "you are more important than I am." Jack expands the consequences: "The loss of self coincides with a loss of *voice* in relationship. Voice is an indicator of self. Speaking one's feelings and thoughts is part of creating, maintaining, and recreating one's authentic self."[16]

The silence we have been advocating is not an excuse for laziness, shyness, hiding behind a socially acceptable persona, or never taking an unconventional position. On the contrary—remembering as before, as Bachelard suggests, that silence leads us back into the world: "Where is our first suffering? We have hesitated to say . . . . It was born in the hours when we have hoarded within us things left unsaid. Even so, the stream will teach you to speak; in spite of the pain and the memories, it will teach you euphoria through euphuism, energy through poems. Not a moment will pass without repeating some lovely round word that rolls over the stones."[17]

CHAPTER NINE

GASTON BACHELARD AND CHILDHOOD EDUCATION

Our childhood bears witness to the childhood of man, of the being touched by the glory of living.

Bachelard does not write directly about how to educate a child. He does advise adults to re-look at the world that we take for granted, to see it with the openness of the child, to allow ourselves to experience wonder. He speaks of retaining in adulthood all the qualities of receptiveness, the freshness as if seeing beauty for the first time. In *The Poetics of Reverie*, the chapter "Reveries Toward Childhood" is almost entirely devoted to the *remembrance* of childhood, the rapturous memories of childhood.

Bachelard himself must have had a joyful experience growing up. He wants to remind us how, through our reveries and daydreams, we can return to that imaginary space of bliss and wonder: "This beauty is within us, at the bottom of memory . . . . When we go looking for it in our reveries, we relive it even more in its possibilities than in its reality. We dream of everything that it could have been; we dream at the frontier between history and legend."[1]

No guide is better than Bachelard for leading us gently but firmly into that forgotten land when he says: "The *permanent child* alone can return the fabulous world to us." Reveries that carry us back toward our childhood have such attraction, such soul quality. In his words: "The reason for this quality (the soul quality) which resists the experiences of life is that childhood remains within us a principle of deep life, of life always in harmony with the possibilities of new beginnings."[2]

As we have seen, Bachelard calls attention to the appeal of the classical elements—earth, air, fire, and water. By staying in tune with the elements, we retain a freshness, an alertness, an openness to joy. For me, this approach has something in common with the English poet Gerard Manley Hopkins's description of the "freshest, deep down things." We are reminded of how love first sprang forth in our hearts: "We cannot love water, fire, the tree without putting love into them, a friendship that goes back to our childhood . . . . We love them with a new found childhood, in a childhood which is latent in each of us."[3] We discover in the depths, "the archetype of simple happiness." Once again, Bachelard is an original. He is saying that because we respond to the material world, we learn what love is about. The things of the world provoke love.

Bachelard, always an advocate of activities that bring joy and relieve anguish, suggests returning as an adult to some of the activities of childhood—for him this means looking at the world with attentive, accepting, credulous eyes, neither passing by without glancing at the buoyant life around us nor approaching every adventure with skepticism. "Our childhood bears witness to the childhood of man, of the being touched by the glory of living." Later in life, part of the joy we experience in reading a certain poem or looking at a painting or sculpture is in being caught up in the writer's view or the artist's eye which makes us aware of fresh details in the world. "Anguish is artificial: we were meant to breathe freely," which is reason enough to follow his suggestion.

The world of matter is the best teacher we have. Speaking of the earth, he says, "Rocks teach us the language of hardness." Speaking of water, the stream causes us to hear better and to speak better. Bachelard explains: "The stream will teach you to speak . . . . Not a moment will pass without repeating some lovely round word that rolls over the stones." Speaking of the vital thing that we learn from the air surrounding us, we learn to hold our heads high, to live "vertically," or in Bachelardian terms, "a courage to live in opposition to weight." From the flame too, we learn to aspire to greater heights. One who "learns the lesson of the flame, realizes that he must right himself. He rediscovers the will to burn high."[4] Fire is the ultra-living element.

Besides a more attentive regard for the world and taking time to relive simple pleasure in daydream, Bachelard urges us to partake in activities that bring us in closer contact with the elements. Making a batter and baking it, or using clay and water to mold something precious brings unexpected joy: "Sculpting clay!—a childhood dream, a dream that returns us to our childhood. As it is often said, for a child all things are possible. As children we were all painters, ceramists, botanists, sculptors, architects, hunters, and explorers. What has become of all that?"[5] One of the reasons that we, as teachers or as parents or grandparents, feel revived by being around children is their attachment to the vivaciousness of life. One of the reasons must be that, as Bachelard remarks in *Earth and Reveries of Will*: "A child let be will mold a chicken or a rabbit, creating life." The growing child is porous in ways that get closed off or shut down later. We acquire defenses and adopt skeptical attitudes in maturing. The child is naturally attuned to nature and absorbs the materiality of the world around: "The child's reverie is a materialistic reverie. The child is a born materialist. His first dreams are dreams of organic substances," Bachelard argues in *Water and Dreams*. This Chinese poem expresses it well:

> It is the child that sees the primordial
> > Secret in Nature
> And it is the child in ourselves that we
> > return to.
> The child within us is simple and daring
> > enough to live the Secret.
> > — Chuang Tzu

What Bachelard does so well, where his special uniqueness lies, is his detailed exploration of how our encounters and our interactions with the world reveal our innermost identity, for: "Matter is a revelation of being." It seems obvious when he says it, but no one ever was quite so specific. It is in these engagements that we acquire our sense of an authentic self. Childhood is an extended time of exploration—of touching, of feeling, of smelling, even of tasting matter. These propensities are not just acts of idle curiosity. An exploring child really wants to know what is out there and how different it is from his or her own body.

This contact establishes boundaries and directions, not only physical. Everything that a child touches teaches the ego how to deal with the world. The touch sense encloses us within the boundary of the body while distinguishing inside from outside, delineating what is me from what is you. The developing kinesthetic sense delineates space for us.

Ultimately, from our many encounters with the world of matter, destiny begins to drag us in one direction instead of another. Thus, a life acquires impetus. Our choices become more focused. Our energies and will are thereby engaged. Matter, therefore, really matters.

Bachelard's favorite element was water, even though he wrote three books on elemental fire. Here are some of his observations: "Water and warmth are the two things vital to our well-being. We must know how to be economical with them." "Water draws to itself all images of purity." As with every element, there are natural qualities that we may have taken for granted. And then, by extension, there are the subtle qualities that the imagination adds, such as: "One drop of powerful water suffices to create a world to dissolve the night."[6] Another of the observations I like is: "Dew is a morning substance, the *distillate of dawn*." This statement, of course, occurs in *Earth and Reveries of Will* in a chapter on dew and pearls. Every time I think I can read all his thoughts on water in the primary book related to that element, *Water and Dreams*, I discover water reappearing in all the other books. Bachelard often speaks of water as the necessary quality of flow and rhythm needed in poetry: "Water is the mistress of liquid language, of smooth flowing language, of continued and continuing language, of language that softens rhythm and gives a uniform substance to differing rhythms."[7]

It is understandable why Bachelard was always attracted to poets and poetry. They, like he, observe the material world with careful, loving eyes. Rainer Maria Rilke wrote in his letters: "The poet hates approximations." As did Bachelard. The scientist in him may have taught him to observe details accurately, but, for him, imagination supplied the joy and dynamism in all encounters. In *Earth and Reveries of Will*, the imagination is an "accelerator of the psyche."

In *The Poetics of Reverie,* Bachelard calls poetry beneficial and "the summit of aesthetic joys." The reason for this is that the *quality of childhood*, which the soul is never deaf to, "is communicable . . . . By the poet's grace we have become the pure and simple subject of the verb 'to marvel.'"[8]

Finally, in *Earth and Reveries of Will,* Bachelard provides us with a clear directive for both our own lives and for our responsibilities to the children in our care: "We must return, in innocent wonder, to things themselves."

CHAPTER TEN

THE ANGELIC IMAGINATION

We take part in imaginary ascension because of a vital need, a vital conquest as it were, of the void.

Bachelard had such a caring attitude toward the earth, the sky, the sea, and equally toward humanity with all its foibles, that the question often arises: "Did he have a religious bent?" After all, his academic grounding was in science, a discipline often considered somewhat indifferent, if not antagonistic, to religion. But, as we have illustrated throughout this book, he identified with and loved matter and the things of the world.

In the realm of the empyrean, Bachelard was a non-conformist. His angels bear no theological imprint. He has been called, perhaps not unfairly, a pagan. As a twentieth-century devotee of the history of science, he was concerned primarily with images of elemental matter. When in mid-career his philosophical focus shifted away from the exclusively empirical, he began a search into the imagination and will, those two quintessential human faculties, which absorbed much of his later career. His explorations convinced him that will and imagination, even when misguided in application, even when caught up in carnality, bear a quotidian relationship to the aspiring, higher side of human nature, that part which yearns for metaphysical meaning. This connection is a linkage, however tenuous, between the pragmatic Bachelard and the metaphysial fraternity of angels. The old adage applies: "Angels can fly because they take themselves lightly."

To Bachelard, the chief characteristic of imagination is mobility; and will, far from being a static faculty, enters into and empowers any action. The definition of will as something akin to ego does not satisfy him. He insists that will includes an almost instinctual responsiveness to challenge and a characteristic surge of psychic energy. The loci of one's will and imagination provide fundamental insights into personal character and determine the teleological direction of one's life. This is a vitalist doctrine which he espouses, suggesting that life processes are determined by the dynamics of the interaction of will and imagination more than by purely mechanical means.

This coupling of foundational human characteristics to angelhood, in becoming more than a mere polemic outreach, is a concomitant of his basic philosophy. Indeed, Bachelard sees angel images as pervasive in art, literature, philosophy, and theology, and therefore as integral to our cultural inheritance. He finds their appeal revealing. An angel's wing, like a bird's, symbolically exemplifies human desire for mobility, a transcendence of the earth-bound. Angels shuttle back and forth between the natural and the supernatural, ethereal messengers echoing the human yearning to bridge two realms. Their invisible presence is more powerful and numinous to the imagination than the visible. They are part of Bachelard's world of "the unreal" through which the real is delineated, the unconscious through which consciousness is grounded, the invisible through which the visible is defined. Even the vernacular of today—"winging it," "lightening up"—brings into metaphorical play what Bachelard describes as humanity's ever-present upward yearning. It is a yearning based on human need, the need to break Gulliver's threads, to soar free like angels.

Through imaginative borrowings from Swedenborg, Balzac, and Villiers de L'Isle-Adam, Bachelard builds his pastiche of literary images. A soaring wing opposes the downward drag of gravity. Luminous, astral light reflects transparency and inner insight. He exploits even the ubiquitous quality of the angel: it unites androgynously the best of the male and the female.

Always, images play upon the human will, galvanizing us into action beyond mere conscious acts of volition, forcing us into interaction with the world, driving us to be masters of it, ultimately to transcend it. Bachelard ascribes to the earth a will of its own, one in continuing contest with the human will. Similarly, the natural world is imbued with both soul, *anima mundi*, and spirit, *spiritus mundi*, which challenge the individual soul and spirit. This titanic competition whets humankind's thirst for improvement, moving on an upward trajectory, along the path of the angels.

In *The Right to Dream*, Bachelard emphasizes that to become effectively spirit, humankind must be "a will straining toward its destiny; must be a will to youth, a will to regeneration."[1] It is straining toward a never-to-be reached goal which he calls "superhuman becoming." In psychology we relate this urge toward youth and regeneration to the *Puer (or Puella) Aeturnus* complex exemplified by Eros, who in the late Greek pantheon is a daimon with both dark and bright angelic qualities. Eros brings desire, or perceived need, into human lives, thereby fusing the longing urge. But the son of Aphrodite struggles with maturity and the need to accept responsibilities for his actions. He yearns to remain free, unshackled, able to fly home to the security of mother at the first sign of trouble from Psyche. The dynamic action epitomizes the awakening between love and the soul in search of liberation, probing new boundaries, and aspiring always to transcend them.

Bachelard implies that the pursuit of transcendence is a fundamental human urge that moves us *pari passu* with the angels. In doing so he differentiates between the will to verticality and the will to the horizontal, which shackles us to the purely practical. The urge to ascend or descend on an axis of verticality governs our moral judgments in the use of spatial metaphors to describe the reach upward to a higher nature or downward to a deepening of the dimensions of personality. In *Air and Dreams,* Bachelard explains the principle of ascensional imagination:

> Of all metaphors, metaphors of height, elevation, depth, sinking, and the fall are the axiomatic metaphors par excellence . . . . They govern the dialectic of enthusiasm and anguish . . . . It is impossible to express moral values without reference to the vertical axis. When we better understand the importance of a physics of

poetry and a physics of ethics, then we will be
closer to the conviction that every valorization is
a verticalization.[2]

The ascensional imagination is always waiting to be
activated. When we climb a mountain, upward desire, the
will to ascension impels us: "The most physical kind of
ascension is thus a preparation for the final assumption."[3]
The thrill of attainment commingles with fear of the
fall, or the fear of the abyss: "We take part in imaginary
ascension because of a vital need, a vital conquest as it
were, of the void."[4]

Typically in a dream when we rise in imaginary flight,
the abyss which before felt as if it were grabbing at our heels
begins to lose its terror. Bachelard explains the dynamics:

> Let us for a moment experience dynamically
> this dominance over the chasm: we notice the
> abyss loses its distinctive features because we
> move away from it. One who ascends sees the
> obliteration of the abyss' characteristics. For him
> the abyss dissolves, becomes hazy, and grows
> more obscure.[5]

Bachelard contends that in our dreams we never
imagine ourselves with angel-like wings until we have
already mastered the art of flying: "Anyone who flies feels
that he has wings when he need no longer make an effort to
fly."[6] Slyly perverse, he suggests that wings on the heels are
more effective than wings on the shoulders and that even
the god Hermes has to use pediform attachments when off
on an aerial jaunt. He observes that Michelangelo has only

to lift the foot of one of his painted seraphs to suggest the ability to fly and that artists generally portray angels as if they were swimming in the firmament. Somewhat disdainfully, he notes that Leonardo da Vinci painted only mechanical, exterior wings, whereas Dante has no need for such props; his angels fly without the specific image of wings. He can render interior wings.[7]

It is in his preface to Honore de Balzac's *Séraphîta* (1955) that Bachelard specifically discusses his approach to angelology. Here he underlines Balzac's direct and indirect borrowings from Emanuel Swedenborg, the eighteenth-century Swedish philosopher who speaks to all those who respond to the angelic call. Commenting on *Séraphîta,* he says:

> A couple of pages may be enough to endorse the innate Swedenborgian enlightenment, to lend confidence to those direct, native impulses that launch human nature on an unwaveringly vertical course. Balzac had within him, as "engram" of all imaginary ascension, that Swedenborgian dynamism.[8]

It is the "unwaveringly vertical" urge that unites Balzac to Swedenborg, Bachelard to them both, and humanity with the angels.

Bachelard, along with Balzac, finds Swedenborg's angelic depictions fascinating. In an absorbing chapter of *Heaven and Hell,* Swedenborg asserts that "Each Angel Has a Perfect Human Form," eyes, ears, nose, etc. Each is a single person and a part of heaven: "Man, too, to the extent that he accepts heaven is also a recipient entity, a heaven and an

137

angel."[9] Only blind ignorance sustains the belief that angels are incorporeal, airy creatures with intellects but without form. He finds intellectuals and the clergy particularly guilty of this skepticism; simple people cannot be so easily misled, since they could never comprehend anything that had no form: "That is why the angels carved or painted in churches are invariably represented as people."[10] To Bachelard, Swedenborg's vision "'solidified' the creatures of heaven" by depicting the dynamics of vertical ascension, by sensitizing the reader "to approach each image in its quality as departure, as magnet for ascension, as standing invitation to an aerial future."[11]

Unlike Bachelard, Swedenborg claims through inner vision, through his spiritual eyes, actually to have seen angels. He writes:

> This is in the spiritual world, while everything physical is in the natural world. Like sees like because it is made of like material. Further, the body's organ of sight, the eye, is so crude that it cannot even see the smaller elements of nature except through a lens, as everyone knows. So it is still less able to see things above the realm of nature like the things in the spiritual world. However, these things are visible to man when he is withdrawn from physical sight and his spirit's sight is opened . . . . It is how I have seen angels.[12]

Swedenborg's angels are in a state of radical innocence: "What is true cannot be bonded to what is good, or vice versa, except by means of innocence."[13] Innocence is the acceptance that all knowledge is derivative and the

138

willingness to be led like a child by the Lord. This finds an echoing, non-theological chord in Bachelard, who stresses opening our eyes to see the world with a child's awe and innocence, not with the naiveté of inexperience—as with Dostoevsky's Prince Myshkin, for example—but rather as a celebration of the earth's wonders. It is an innocence that is beyond ego, a post-heroic mode. There are no tasks, no goals, no work to accomplish, only the acceptance of the paradisiacal state of mind. It results from aspiring to the good, the true, the beautiful, and discovering it within oneself as well as externally in the world.

Bachelard finds the nearest approach to angelhood in human form in Balzac's novel, which he describes as "a poem of will, a dynamic poem"[14] and a "meditation upon the regeneration of human nature by the Will."[15] Unlike Eastern meditation on being, it is the emphasis on "supernatural becoming," or "the human being's destiny of transcendence"[16] that is the central theme of *Séraphîta*. Notice the play on the name Seraphitus/Séraphîta as the androgynous near-angel protagonist. Séraphîta, the embodiment of the perfected creature, has no world-saving mission to accomplish. She has male and female qualities so complementary as to constitute near self-sufficiency—a divine amalgam of the sexes contoured in a human form. As a symbolic bridge between the sexes, she is viewed through different prisms—as a male by young Minna but as a female by the male characters. She is also a link between mortal and spiritual worlds, the synthesis in Bachelard's view "of earthly being and immortal being."[17] Thus she approaches the state of androgynous angelhood in mortal form. Bachelard's emphasis suggests by implication the ultimate possibility of attaining perfectibility in the human realm.

Bachelard invariably uses epigrams to emphasize his main points. In his preface to *Séraphîta* he augments his theme with an epigram from Villiers de L'Isle-Adam's *Axel:*

> Fulfill yourself in your astral light! Come forth!
> Reap! Arise! Become your own flower![18]

To fulfill one's being with astral light means fulfilling one's spiritual potential. "Superhuman becoming" is the soul's straining to become spirit. Success depends on the intensity of this urge toward the ascensional, toward angelhood. Bachelard extols the virtue of Dante as "the poet most attracted to verticality."[19] He likens this quality to the same force within Balzac and Swedenborg, "the profound traces of which Balzac urges the sympathetic reader to discover within himself."[20] Always we can resonate with the imagination of a poet. Thus it is through a new definition of, a redefining of *participation mystique*, through participation in the "psychic lyricism" of the poet, that we can join with imagination and will in the soaring significance of the angels.

EPILOGUE

When Bachelard died in October 1962, it marked the end of a highly productive life. Between 1928 and 1961 he had written twenty-three books—twelve on the philosophy of modern science, two on time and consciousness, nine on poetic imagination, and a tenth book that he left unfinished that has been translated by the Dallas Institute, *Fragments of a Poetics of Fire*. He was best known and quite famous for his books on the philosophy of science when, in 1938, he wrote the surprising book *The Psychoanalysis of Fire*. Who could ever have guessed that the man who had opposed the "taciturn scientific mind" to the "expansive poetic mind" would launch into such a thorough exploration of the "material imagination"? If the audacity of making an element, such as fire, the patient in the examination room is hard to imagine now, how very strange it must have seemed then. Freudian psychology was just coming into being and arguments raged concerning whether it was a science or part of the field of humanities. In Bachelard's subsequent books, he turned toward Jungian psychology and phenomenology, borrowing the method of phenomenology to examine the physical world and its effect on human personality. He shared with Jung an interest not in the abstract theories but in the depth of psychological awareness. In fact, he complained about many of the psychoanalytic concepts, which he felt left the human soul out of the equation. He cared intensely for what it means to be fully human and helped to return the fulsomeness of the human mind and heart to center stage. He was a "subversive humanist," as Mary McAllester Jones calls him in her 1991 book. He was engaged with his explorations on imagination until he died. In his final manuscript, in a note on the side of a page, he wrote: "The more I work, the more diverse I become."

My thirty-year career as Director of Publications at the Dallas Institute, spearheading the Institute's Bachelard Translation Series, has instilled in me his sense of wonder. Mark di Suvero, the renowned sculptor, also moved by Bachelard's sense of wonder, once sent me a signed print and expressed his gratitude for "doing so much to keep the poetry/flame/imagination of G[aston] Bachelard glowing." I have become convinced that it is a crime to let slip by unappreciated or unnoticed any moment of awareness of the earth's beauty. Without being too rhapsodic, I can only wish that Bachelard's unique gift of enlarging the scope of imagination will awaken in you a similar elation and joy.

APPENDIX

In an essay on the significance of Bachelard's work for the theory and practice of astrology and other forms of divination, Jean Hinson Lall provides some reflections on the philosopher's natal horoscope (Figure 1). She made these telling remarks about it, as well as the reasons why we might want to study the natal horoscope of a philosopher or any other thinker:

> Jung commented that he regarded his own contribution to his field as "subjective confession," suggesting that a theory is an expression of the theorist's individual character, ancestry, and disposition. Astrologically, we could say that a person's thought is not merely subjective and confessional, but an expression of the objective, archetypal ground of his or her existence, which shows itself in the symbolic map of the cosmos as viewed from the time and place of birth. Reflection on the natal horoscope of a scholar, writer, or artist helps us to appreciate what the individual brings that is unique, its brilliance as well as its limitations. It is not that the philosopher's horoscope explains

his thought; still less does it explain it away in terms of personal complexes.[1]

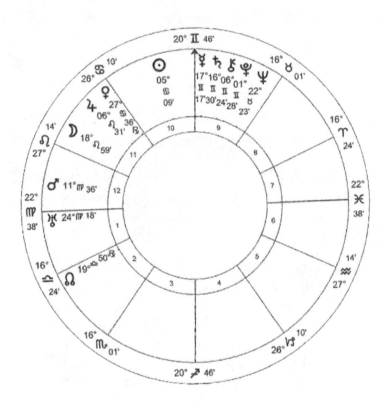

Figure 1

Natal Horoscope of Gaston Bachelard: 26 June 1884, 11:00 am LMT, Bar-sur-Aube, France; 48°N14', 04°E43'; Tropical zodiac, Geocentric, Topocentric Houses. Reprinted with the permission of Angela Voss and Lall, editors, *The Imaginal Cosmos: Astrology, Divination, and the Sacred* (Canterbury: The University of Kent, 2007), 119.

APPENDIX

A student of mine, Betty Regard, tasked with interpreting Bachelard's horoscope once wrote this: "All four elements are strong in his horoscope. Elevated sun in the 10th house—will do well in his profession, [garner] recognition from other people. The moon is in Leo—a fire sign—[a sign of] feeling and emotion, expressed in a dynamic way, even in science. 1st House in Virgo, an Earth sign: work, study, and teach. On the cusp of the 1st House Uranus—originality, eccentricity, genius. He has four planets in the 9th House Air—close to the tenth. Mercury conjunct Saturn—higher education."

Jean Lall, more professionally and thoroughly, suggests this detailed interpretation, which seems to describe aptly Bachelard's range of interests from the philosophical and scientific to the personal and the near-at-hand:

> In Bachelard's map the Sun, Moon, and planets are clustered in a wedge covering just over one third of the zodiac (122 degrees), suggesting a highly focused personality. The area of concentration covers primarily the ninth through twelfth houses, with Uranus lying just inside the first house, indicating that the emphasis in the person's life is likely to be on collective and impersonal questions (these houses being philosophical, professional, scientific, humanitarian, global, and mystical in orientation). The planets and the ascendant, however, are all in spring and summer signs. Sun and Venus in Cancer, Mercury and Saturn in Gemini, and Moon and Jupiter in Leo emphasize the personal and near-at-hand, the

145

creative, one's own self-expression. Gemini and Virgo want to learn, teach and communicate at a practical, hands-on level. These early signs bring a certain freshness, warmth, and joy to the sober concerns of the late houses, while being placed in the late houses gives the early-sign planets an impersonal perspective on personal matters and on human existence itself.[2]

She further adds:

The Sun in the tenth house typically shows a drive toward professional achievement and public recognition, a tendency to rise to the top through a combination of disciplined effort, talent and charisma . . . . The Sun represents the radiant energy of the heart, and the sign in which it is placed indicates the nature of the fire that burns in the person's inner hearth . . . . The placement of Bachelard's Sun in the Moon-ruled water sign Cancer helps us to see why, despite his strong scientific bent, water was his favourite element, and why he was so comfortable with flux, with the feminine and the maternal, and with poetic language.[3]

Lall's acute observations seem pertinent to me in fully appreciating Bachelard's talents and his contribution to human knowledge.

ACKNOWLEDGEMENTS

As always, writing a book constitutes a combining of multiple efforts and skills. This one is no different. It will be difficult to give to each contributor the proper due. I must first profusely thank Robert Sardello for having the idea of gathering together my Bachelard lectures and papers over many decades of teaching. I thank him too for his splendid introduction. And then I must add Robert and Cheryl Sanders-Sardello's classes in Spiritual Psychology, which provided the opportunity to consider multiple sides of Bachelard's gifts. Especially true is Chapter Four, "Alchemy, The Senses, and Imagination," given at one of their seminars at the School of Spiritual Psychology, and Chapter Ten, "The Angelic Imagination," which was presented at the Dallas Institute of Humanities and Culture, and produced in 1995 a book published by Continuum Press on the Angels.

Richard Lewis, in his lifelong work with artistic projects and public school children in New York City, through his organization, The Touchstone Center for Children, and in his many books with Touchstone Center Publications, is in the trenches actively proving Bachelard's theory about the centrality of the imagination. Richard has contributed to our conferences on Bachelard and invited me to speak in 2005 at Bank Street College in New York

City, which resulted in Chapter Nine, "Gaston Bachelard and Childhood Education."

I must add a digression here to tell you how Richard contributed so much to the entire saga. In the early 1960s he was invited to visit Howard Greenfield, then the publisher and editor of Orion Press, for a weekend in Connecticut. Greenfield presented Richard with a manuscript to read and asked him to give, at the end of the weekend, his opinion about whether Orion should publish the book or not. Richard was captivated by the book and said, by all means, an affirmative "yes." You may have guessed by now that this book, *The Poetics of Space,* published in 1964, was the very one that launched the first translation of Gaston Bachelard in English in the U.S. and appealed to so many artists and architects and also sealed my destiny in graduate school.

Any of you who are devoted to the writings of the late James Hillman (co-published by the Dallas Institute Press) know of his use of many references to Bachelard. In 1980, Jim, along with Drs. Louise and Don Cowan, Robert Sardello, Gail Thomas, and myself, was one of our six Founding Fellows of the Institute, and always remained a dear, dear friend. His wife Margo McLean's painting appropriately graces the cover of this book.

Through Robert Sardello, I met Lee Nichol, a superb editor who always kept a rein on my ability to extend digressions endlessly. Lee knows how to forge a whole, seamless pattern out of disparate and random thoughts and knows when enough is enough. Sarah Theobald-Hall worked meticulously through verifying all the references, but much more, her literary sensibilities saved the day on many occasions.

ACKNOWLEDGEMENTS

I am lucky to have three adult and talented children: my sons, Ethan and Eric, have superb financially oriented talents. Personally, both have always recognized how vital teaching and writing efforts are for me. It is my daughter Natasha, with her sense of style and acute awareness of words, who is an accomplice in all my writings. I could never commit any work to paper (or digital medium) without her perusal. My late husband, Kenneth Bilby, with his journalistic background, was the finest editor I have ever known. Getting his approval for a chapter, for a paragraph even, made my day. Janie Milburn, my assistant of twenty-nine years, provided countless hours of administrative attention.

At the Dallas Institute are many supportive friends, but especially Larry Allums, the Director, and Claudia Allums, Head of the Teacher's Institute—both have encouraged me to teach classes in Bachelard's works and furthered the publications of our Fellows. I will always be grateful to Robert Scott Dupree for his thirty years of continuous engagement in the complicated art of translation. Without his extensive knowledge of French and his grasp of lyric language, so much a part of Bachelard, our seven translations from French to English would have been impossible. He has had, more than once, to step in and completely revise our renditions in the Bachelard Series. Mary McAllester Jones, of all translators in the Series, was most able to catch the wisdom, the originality, uniqueness, together with Bachelard's lyrical voice. Finally, Suzanna Brown and I have shared the pleasures and pains of publishing now for many years. She is one of the most talented of graphic designers, able to mesh her originality with whatever projects we tackle in this hermetic endeavor.

WORKS CITED

Audisio, Gabriel. "L'Age de Pierre" [The Stone Age], *Poemes du lustre noir*. Marseilles: Robert Laffont, 1944.

Bachelard, Gaston. *Air and Dreams : An Essay on the Imagination of Movement*, trans. Edith R. Farrell and C. Frederick Farrell. Dallas: The Dallas Institute Publications, 1988.

―――. *Earth and Reveries of Repose: An Essay on Images and Interiority*, trans. Mary McAllester Jones. Dallas: The Dallas Institute Publications, 2011.

―――. *Earth and Reveries of Will: An Essay on the Imagination of Matter,* ed. Joanne H. Stroud, trans. Kenneth Haltmann. Dallas: The Dallas Institute Publications, 2001.

―――. *The Flame of a Candle*, ed. Joanne H. Stroud, trans. Joni Caldwell. Dallas: The Dallas Institute Publications, 1988.

―――. *Fragments of A Poetics of Fire*, ed. Suzanne Bachelard, trans. Kenneth Haltman. Dallas: The Dallas Insitute Publications, 1990.

―――. *On Poetic Imagination and Reverie*, trans. Colette Gaudin. New York: Bobbs-Merrill, 1971.

―――. *The Poetics of Reverie: Childhood, Language, and the Cosmos*, trans. Daniel Russell. Boston: Beacon Press, 1969.

Works Cited

————. *The Poetics of Space,* trans. Maria Jolas, foreword by John R. Stilgoe. Boston: Beacon Press, 1994 (also 1964 Beacon Press edition).

————. *La Philosophie du non: essai d'une philosophie du nouvel spirit scientifique.* Paris: Presses Universitaires de France, 1940.

————. *Psychoanalysis of Fire.* Boston: Beacon Press, 1964.

————. *The Right to Dream,* ed. Joanne H. Stroud, trans. J. A. Underwood. Dallas: The Dallas Institute Publications, 1988.

————. *Water and Dreams: An Essay on the Imagination of Matter,* trans. Edith R. Farrell. Dallas: The Dallas Institute Publications, 1999.

Berry, Patricia. *Echo's Subtle Body: Contributions to An Archetypal Psychology.* Dallas: Spring Publications, 1987.

Berry, Wendell. "Prayers and Sayings of the Mad Farmer." *The Mad Farmer Poems.* Berkeley: Counterpoint, 2008.

Breton, André. *Surrealism and Painting,* trans. Simon Watson Taylor. Boston: MFA, 2002.

Bruner, Jerome. *On Knowing: Essays for the Left Hand.* Cambridge: President and Fellows of Harvard College, 1997.

Burckhardt, Titus. *Alchemy: Science of the Cosmos, Science of the Soul.* Baltimore: Penguin Books, 1972.

Casey, Edward. "Anima Loci." *Sphinx: A Journal for Archetypal Psychology and the Arts,* 5, 1993.

Caws, Mary Ann. *Surrealism and the Literary Imagination: A Study of Breton and Bachelard*. The Hague: Mouton, 1966.

Caws, Peter. *Yorick's World: Science and the Knowing Subject*. Berkeley: University of California Press, 1993.

de Chardin, Teilhard. *The Divine Milieu*, ed. Bernard Wall. London: William Collins Sons & Co., 1960.

Christofides, C. G. "Bachelard's Aesthetics." *The Journal of Aesthetics and Art Criticism*, 20, No. 3, 1962.

———. "Gaston Bachelard and the Imagination of Matter." *Revue Internationale de Philosophie* 66. University of Belgium, 1963.

Chopra, Deepak. *Ageless Body, Timeless Mind*. New York: Harmony Books, 1993.

Cowan, Donald. *Unbinding Prometheus: Education for the Coming Age*. Dallas: The Dallas Institute Publications, 1988.

Eliade, Mircea. *Sacred and the Profane: The Nature of Religion*, trans. Willard R. Trask. New York: Harcourt, Inc., 1987 (also 1959 edition).

Farrell, Edith. "Introduction." *Water and Dreams*, Gaston Bachelard, trans. Edith Farrell. Michigan: University Microfilms, 1965.

Fox, Matthew. *The Coming of the Cosmic Christ: The Healing of Mother Earth and the Birth of a Global Renaissance*. San Francisco: Harper Collins, 1988.

Works Cited

Gaudin, Colette. "Preface." *On Poetic Imagination and Reverie*, Gaston Bachelard, trans. and ed. Collette Gaudin. Dallas: Spring Publications, Inc., 1987.

Gilson, Etienne, "Foreword." *The Poetics of Space*, Gaston Bachelard. Boston: Beacon Press, 1964.

Hauptmann, Gerhardt. *La Cloche engloutie; conte dramatique en cinq actes*, trans. A. Ferdinand Herold. Paris: Societe du Mercure de France, 1897, III.

Hillman, James. *Alchemical Psychology*, Volume Five of *Uniform Edition of the Writings of James Hillman*. Putnam, CT: Spring Publications Inc., 2010.

Hillman, James. *Archetypal Psychology,* Volume One of *Uniform Edition of the Writings of James Hillman*. Putnam, CT: Spring Publications, Inc., 2013.

Jack, Dana Crowley. *Silencing the Self: Women and Depression*. Cambridge and London: Harvard University Press, 1991.

Jones, Mary McAllester. *Gaston Bachelard, Subversive Humanist: Texts and Readings*. Madison: University of Wisconsin Press, 1991.

————. *The Philosophy and Poetics of Gaston Bachelard*. Washington D.C.: Center for Advanced Research in Phenomenology and University Press of America, 1989.

Jung, C. G. *The Archetypes and the Collective Unconscious*, trans. and ed. G. Adler and R. F. C. Hull, vol. 9, bk. 1, *Collected Works of C. G. Jung*. Princeton: Princeton University Press, 1975.

————. *Civilization in Transition,The Collected Works of C. G. Jung.* Trans. and ed. G.Adler and R.F.C. Hull, Vol. 10, Bollingen Series XX. Princeton: Princeton University Press, 1970.

————. *Psychologie und Alchemie.* Zurich: Rascher, 1944.
Lall, Jean Hinson. "Watering the Roots of Astrological Theory and Practice: Gaston Bachelard's Contribution to a Philosophy of Divination." *The Imaginal Cosmos: Astrology, Divination, and the Sacred*, ed. Angela Voss and Jean Lall. Canterbury: The University of Kent, 2007.

LeRoy, S. J., Pierre. "Teilhard de Chardin, the Man." *The Divine Milieu*, Teilhard de Chardin, ed. Bernard Wall. London: William Collins Sons & Co., 1960.

Lipson, Michael. *The Stairway of Surprise: Six Steps to a Creative Life.* Great Barrington, MA: Anthroposophic Press, 2002.

McCraty, Rollin, Raymond Trevor Bradley and Dana Tomasino, "The Resonant Heart." *Shift: At the Frontiers of Consciousness*, 5. December 2004-February 2005.

Ménard, Louis, trans. *The Hermetica of Hermes Trimégiste.* Paris, 1866.

Michell, John. *The Earth Spirit: Its Ways, Shrines and Mysteries.* New York: Thames and Hudson, 1975.

Papineau, David. "Get a Grip." Review of *The Hand: How Its Use Shapes the Brain, Language, Culture*, Frank R. Wilson. *New York Times Book Review*, July 19, 1998.

Paracelsus, Theophrastus. *Selected Writings*, ed. Jolande Jacobi, trans. Norbert Guterman. Princeton: Princeton University Press, 1958.

WORKS CITED

Parks, Gordon. *Arias In Silence*. Boston, New York, Toronto, London: Bulfinch Press, 1994.

Picard, Max. *The World of Silence,* trans. Stanley Godman. South Bend, Indiana: Gateway, 1952.
Pire, François. *De l'imagination poetique dans l'oeuvre de Gaston Bachelard*. Paris: Librairie José Corti, 1967.

de Quincey, Christian. *Radical Nature: The Soul of Matter.* Rochester, VT: Park Street Press, 2002.

Steiner, George. *Language and Silence: Essays on Language, Literature and the Inhuman.* New York: Antheneum, 1977.

Storr, Anthony. *Solitude A Return to the Self.* New York: The Free Press, 1988.

Stroud, Joanne H. *The Bonding of Will and Desire.* New York: Continuum Publishing Company, 1994.

Swedenborg, Emanuel. *Heaven and Hell*, trans. George F. Dole. West Chester, PA: Swendenborg Foundation Publishers, 2000.

Waley, Arthur. *The Way and its Power: A Study of the Tao Te Ching.* London: 1942.

Wilson, Frank. *The Hand: How Its Use Shapes the Brain, Language, and Culture.* New York: Pantheon Books, 1998.

Yeats, W. B. *The Collected Works of W. B. Yeats, Volume 1: The Poems*, ed. Richard J. Finneran. New York: Scribner, 1997.

ENDNOTES

Preface

1. Bachelard, *The Poetics of Space*, xviii-xix.
2. Ibid., xix.
3. Gilson, "Foreword," in Bachelard, *The Poetics of Space*, xiii.
4. Bachelard, *The Poetics of Space*, xv.
5. Bachelard, *Fragments of a Poetics of Fire*, 11.
6. Jones, *The Philosophy and Poetics of Gaston Bachelard*, 3.
7. Gaudin, "Preface," *On Poetic Imagination and Reverie*, xviii.
8. Bachelard, *Air and Dreams*, book jacket.
9. Breton, *Surrealism and Painting*, 3-4.
10. Caws, Mary Ann, *Surrealism and the Literary Imagination: A Study of Breton and Bachelard*, 5.
11. Jones, *The Philosophy and Poetics of Gaston Bachelard*, 9.
12. Gaudin, "Introduction," in Bachelard, *On Poetic Imagination and Reverie*, xxxi.
13. Jones, *The Philosophy and Poetics of Gaston Bachelard*, 2.
14. Caws, Peter, *Yorick's World: Science and the Knowing Subject*, 274.
15. Lall, "Watering the Roots of Astrological Theory and Practice," *The Imaginal Cosmos*, 119.
16. Hillman, *Archetypal Psychology*, 82.
17. Bruner, *On Knowing: Essays for the Left Hand*, 2-3.
18. Bachelard, *Earth and Reveries of Will*, 261.
19. Bachelard, *The Right to Dream*, 54.
20. Hillman, *Archetypal Psychology*, 56.
21. Ibid., 56.
22. Bachelard, *The Poetics of Space*, 36.
23. Bachelard, *Earth and Reveries of Will*, 260.

24. Pire, *De l'imagination poetique dans l'oeuvre de Gaston Bachelard*, 152.
25. Hillman, *Archetypal Psychology*, 76.

Chapter One: Calling Us to the Things of This World

1. Bachelard, *Earth and Reveries of Repose*, 3.
2. Ibid., 2.
3. Ibid., 202.
4. Bachelard, *Air and Dreams*, 136.
5. Yeats, *The Collected Works, Volume 1: The Poems*, 197, 252.
6. Jung, *Psychologie und Alchemie*, 399.
7. Bachelard, *Earth and Reveries of Repose*, 38.
8. Yeats, *Collected Works, Volume 1*, 55.
9. de Quincey, *Radical Nature: The Soul of Matter*, 251.
10. Qtd. in Pierre LeRoy, S.J.,"Teilhard de Chardin, the Man," in *The Divine Milieu*, Teilhard de Chardin, 13.
11. Bachelard, *The Right to Dream*, 63.

Chapter Two: Why Matter Matters

1. Gaudin, "Introduction," in Bachelard, *On Poetic Imagination and Reverie*, xlvii.
2. Bachelard, *Earth and Reveries of Will*, 17. Unless otherwise noted, all quotations in this chapter are from *Earth and Reveries of Will*.
3. Ibid., 17.
4. Ibid., 38.
5. Bachelard, *Water and Dreams*, 183.
6. Bachelard, *Earth and Reveries of Will*, 38.
7. Bachelard, *The Right to Dream*, 56.
8. Bachelard, *Earth and Reveries of Will*, 97.
9. Bachelard, *The Right to Dream*, 67.

10. Bachelard, *La Philosophie du non: essai d'une philosophie du nouvel spirit scientifique*, 134; in Jones, *Gaston Bachelard, Subversive Humanist*, x.
11. Bachelard, *Water and Dreams*, 10.
12. Bachelard, *The Poetics of Reverie*, 1.
13. Bachelard, *The Poetics of Space*, 28.
14. Bachelard, *Earth and Reveries of Will*, 176.
15. Qtd. in Matthew Fox, *The Coming of the Cosmic Christ*, 19.

Chapter Three: The Seduction of Matter

1. Bachelard, *Earth and Reveries of Will*, 163.
2. Bachelard, *The Poetics of Reverie*, 196.
3. Bachelard, *Water and Dreams*, 19.
4. Bachelard, *The Poetics of Reverie* 25.
5. Bachelard, *Water and Dreams*, 17.
6. Bachelard, *Earth and Reveries of Will*, 247.
7. Ibid., 247.
8. Bachelard, *The Poetics of Space*, xviii.
9. Gaudin, "Introduction," in Bachelard, *On Poetic Imagination and Reverie,* xli.
10. Bachelard, *Air and Dreams*, 3.
11. Bachelard, *Water and Dreams*, 14.
12. C. G. Christofides, "Gaston Bachelard and the Imagination of Matter," *Revue Internationale de Philosophie* 66, 485-486.
13. Edith Farrell, "Introduction," in Bachelard, *Water and Dreams* (University Microfilms edition), 30-31.
14. Bachelard, *Water and Dreams*, 183.
15. Ibid., 91.
16. Ibid., 73.
17. Ibid., 17.
18. Ibid., 18.
19. Ibid., 76.
20. Ibid., 21.
21. Ibid., 20.
22. Ibid., 28.
23. Ibid., 167.
24. Ibid., 15.
25. Ibid., 187.
26. Ibid.,195.
27. Gaudin, "Introduction," in Bachelard, *On Poetic Imagination and Reverie,* liii, f.29.
28. Bachelard, *Earth and Reveries of Will*, 247.
29. Farrell, "Introduction," in Bachelard, *Water and Dreams* (University Microfilms edition), 25-27.
30. Bachelard, *Earth and Reveries of Will*, 135.

31. Ibid., 109.
32. Ibid., 34.
33. Bachelard, *Air and Dreams*, 227.
34. Ibid., 7.
35. Bachelard, *Earth and Reveries of Will*, 19; *The Poetics of Space*, xxxiv.
36. Bachelard, *Water and Dreams*, 27.
37. Bachelard, *The Poetics of Reverie*, 2.
38. Bachelard, *Air and Dreams*, 20.
39. Ibid., 35.
40. Ibid., 1.
41. Ibid., 128-129.
42. Ibid., 131.
43. Ibid., 136.
44. Ibid., 8.
45. Ibid., 14.
46. Stroud, *The Bonding of Will and Desire*, 127.
47. Bachelard, *Fragments of A Poetics of Fire*, xiii.
48. Ibid., 84-85.
49. Ibid., xii.
50. Ibid., 83-84.
51. Bachelard, *Psychoanalysis of Fire,* 12.
52. Bachelard, *Fragments of a Poetics of Fire*, 83.
53. Ibid., 73-74.
54. Ibid., 82.
55. Ibid., 69.
56. Cowan, *Unbinding Prometheus*, 168.
57. Bachelard, *Fragments of a Poetics of Fire,* 99.
58. Bachelard, *Earth and Reveries of Will,* 270.

Chapter Four: Alchemy, the Senses, and Imagination

1. Bachelard, *Air and Dreams*, 11.
2. *The Hermetica of Hermes Trimégiste*, trans. Louis Ménard, 219.
3. Bachelard, *Water and Dreams*, 15.
4. Ibid., 19.
5. Ibid., 14.
6. Bachelard, *Earth and Reveries of Will*, 292.
7. Ibid., 292.
8. Ibid., 292-293.
9. Bachelard, *Water and Dreams*, 20.
10. Lipson, *The Stairway of Surprise*, 61.
11. McCraty, et. al., "The Resonant Heart," *Shift: At the Frontiers of Consciousness*, 5; 15-19.
12. Ibid., 16.
13. Ibid., 17.
14. Jung, *The Archetypes and the Collective Unconscious*, 179.
15. Bachelard, *Earth and Reveries of Will*, 44-45.
16. Ibid., 45.
17. Burckhardt, *Alchemy*, 68.
18. Paracelsus, *Selected Writings*, 21.
19. Burckhardt, *Alchemy*, 189.
20. Hillman, *Alchemical Psychology*, 18.

Chapter Five: Images and Archetypes in Bachelard and Jung

1. Christofides, "Bachelard's Aesthetics," 268.
2. Bachelard, *On Poetic Imagination and Reverie*, 97.
3. Gaudin, "Introduction," in Bachelard, *On Poetic Imagination and Reverie*, xxxviii.
4. Bachelard, *Earth and Reveries of Will*, 8.
5. Jung, *Civilization in Transition*, 64.
6. Ibid., 53.
7. Bachelard, *Earth and Reveries of Will*, 2.

8.   Ibid., 3.
9.   Ibid., 14-15.
10. Ibid., 3.
11. Ibid., 3-4.
12. Bachelard, *Earth and Reveries of Repose*, 38.
13. Bachelard, *Earth and Reveries of Will*, 3.
14. Ibid., 12.
15. Ibid., 10.
16. Ibid., 7.
17. Ibid., 7.
18. Berry, Wendell, "Prayers and Sayings of the Mad Farmer," *The Mad Farmer Poems,* 6.
19. Bachelard, *Earth and Reveries of Will*, 7-8.
20. Ibid., 17.
21. Casey, "Anima Loci," 125.
22. Michell, *The Earth Spirit*, 20.
23. Ibid., 12.
24. Eliade, *Sacred and the Profane,* 13.
25. Berry, Patricia, *Echo's Subtle Body*, 7.
26. Ibid., 2.

Chapter Six: The Dual Imagination of Earth

1.   de Quincey, *Radical Nature: The Soul of Matter*, 251.
2.   Bachelard, *Earth and Reveries of Repose*, 2.
3.   Bachelard, *Earth and Reveries of Will*, 11.
4.   Ibid., 7.
5.   Bachelard, *Earth and Reveries of Repose*, 192.
6.   Ibid., 193.
7.   Ibid., 194.
8.   Bachelard, *Earth and Reveries of Will*, 5.
9.   Bachelard, *Air and Dreams*, 1.
10. Bachelard, *Fragments of a Poetics of Fire*, 4.

11. Bachelard, *Earth and Reveries of Repose*, 40.
12. Bachelard, *Fragments of a Poetics of Fire*, 4.

Chapter Seven: The Hand of Work and Play

1. Gaudin, "Introduction," in Bachelard, *On Poetic Imagination and Reverie,* xli.
2. Papineau, David. "Get a Grip." Review of *The Hand* by Frank R. Wilson. *New York Times Book Review.*
3. Bachelard, *The Right to Dream*, 155.
4. Bachelard, *Earth and Reveries of Will*, 278-279.
5. Ibid., 102.
6. Ibid., 102.
7. Ibid., 109.
8. Ibid., 35.
9. Gaudin, "Introduction," in Bachelard, *On Poetic Imagination and Reverie,* xxxvii.
10. Ibid., xxxvii.
11. Bachelard, *Earth and Reveries of Will*, 109-110. Bachelard quotes Gerhardt Hauptmann, *La Cloche engloutie; conte dramatique en cinq actes*, 145.
12. Ibid., 146. Bachelard quotes Gabriel Audisio, "L'Age de Pierre" [The Stone Age], *Poemes du lustre noir*, 25.
13. Gaudin, in Bachelard, *On Poetic Imagination and Reverie*, 80, f. 10.
14. Ibid., 80, f. 10.
15. Bachelard, *Water and Dreams*, 107.
16. Bachelard, *Earth and Reveries of Will*, 101.
17. Bachelard, *The Right to Dream*, 176.
18. Ibid., 173-174.
19. Yeats, *Collected Works*, 154.

Chapter Eight: The Fullness of Silence

1.   Waley, *The Way and its Power*, 162.
2.   Picard, *The World of Silence*, 67.
3.   Ibid., 17-18, 95.
4.   Ibid., 70.
5.   Ibid., 22.
6.   Bachelard, *The Poetics of Space*, 9.
7.   Bachelard, *Earth and Reveries of Will*, 3.
8.   Bachelard, *Air and Dreams*, 244-245.
9.   Bachelard, *On Poetic Imagination and Reverie*, 24-25.
10. Bachelard, *Earth and Reveries of Will*, 2.
11. Picard, 21.
12. Ibid., 19-20.
13. Ibid., 91.
14. Parks, *Arias In Silence*, 124-125.
15. Jack, *Silencing the Self*, 140.
16. Ibid., 32.
17. Bachelard, *Water and Dreams*, 195.

Chapter Nine: Gaston Bachelard and Childhood Education

1.   Bachelard, *The Poetics of Reverie*, 101. Unless otherwise noted, all quotations in this chapter are from *The Poetics of Reverie*.
2.   Ibid., 124.
3.   Ibid., 126.
4.   Bachelard, *Air and Dreams*, 15; *The Flame of a Candle*, 40.
5.   Bachelard, *Earth and Reveries of Will*, 71.
6.   Bachelard, *Water and Dreams*, 14, 9, 128.
7.   Ibid., 187.
8.   Bachelard, *The Poetics of Reverie*, 127.

Chapter Ten: The Angelic Imagination

1.   Bachelard, *The Right to Dream*, 96-97.
2.   Bachelard, *Air and Dreams*, 10-11.
3.   Bachelard, *The Right to Dream*, 97.
4.   Bachelard, *Air and Dreams*, 59.
5.   Ibid., 60.
6.   Ibid., 58.
7.   Ibid., 54, f.25.
8.   Bachelard, *The Right to Dream*, 95.
9.   Swedenborg, *Heaven and Hell*, 126, 122.
10. Ibid., 124
11. Bachelard, *The Right to Dream*, 96.
12. Swedenborg, *Heaven and Hell*, 125.
13. Ibid., 125.
14. Bachelard, *Air and Dreams*, 58.
15. Bachelard, *The Right to Dream*, 97.
16. Ibid., 94.
17. Ibid., 94.
18. Ibid., 93.
19. Bachelard, *Air and Dreams*, 40.
20. Bachelard, *The Right to Dream*, 98.

ENDNOTES

Appendix

1. Lall, "Watering the Roots of Astrological Theory and Practice," *The Imaginal Cosmos*, 121.
2. Ibid., 121.
3. Ibid., 122.

ABOUT THE AUTHOR

Joanne H. Stroud, Ph.D., taught both literature and psychology at the University of Dallas before becoming a co-founder and Founding Fellow of the Dallas Institute of Humanities and Culture. Currently on the faculty there, she teaches in the Cultural and Spiritual Psychology program. Her book, *The Bonding of Will and Desire*, was published by Continuum in 1994. Thirty years ago she began the task of translating eight books from the works on imagination of French philosopher of science Gaston Bachelard, which culminated in 2012 with the final translation in the series, *Earth and Reveries of Repose*. From her many lectures and classes, this book was born. She has also written *Choose Your Element: Earth, Air, Fire, and Water* (a four-book set), *Time Doesn't Tick Any More*, and *Towers 2 Tall*, with illustrations by Tashasan. All were inspired by Bachelard.